Cells
Tissues
Organs

General Information

Publication data: 'Cells Tissues Organs' is published 12 times a year. Volumes 170–172 with 4 issues each appear in 2002.

Subscription rates: Subscriptions run for a full calendar year. Prices are given per volume.
Personal subscription:
Subscription + postage and handling
CHF 80.– + CHF 19.20 Europe
CHF 80.– + CHF 28.50 Overseas
USD 62.00 + USD 19.20
+ online subscription per volume:
CHF 39.– / USD 30.00
(Must be in the name of, billed to, and paid by an individual. Order must be marked 'personal subscription'.)
Institutional subscription:
Subscription + postage and handling
CHF 834.– + CHF 24.– Europe
CHF 834.– + CHF 35.60 Overseas
USD 641.00 + USD 24.00
+ online subscription per volume:
CHF 58.– / USD 45.00
Airmail postage: CHF 25.20, USD 21.00.
(Extra per volume.)

Microforms: Please direct all enquiries and orders to: University Microfilms Inc.
300 North Zeeb Road, P.O. Box 1346
Ann Arbor, MI 48106-1346 (USA)

Subscription orders: Orders can be placed at agencies, bookstores, directly with the Publisher
S. Karger AG, P.O. Box,
CH–4009 Basel (Switzerland)
or with any of the following distributors:
Deutschland: S. Karger GmbH, Postfach
D–79095 Freiburg
France: Librairie Luginbühl
36, bd de Latour-Maubourg, F–75007 Paris
Great Britain, Ireland: S. Karger AG
4 Rickett Street
London SW6 1RU, UK
USA: S. Karger Publishers, Inc.
26 West Avon Road, P.O. Box 529
Farmington, CT 06085, USA
India, Bangladesh, Sri Lanka:
Panther Publishers Private Ltd.
39, 6th Cross
Wilson Gardens
Bangalore 560 027, India
Japan: Katakura Libri, Inc.
2F, Marushima Bldg, 4-6-17, Yushima,
Bunkyo-ku, Tokyo 113-0034, Japan

Change of address: Both old and new address should be sent to the subscription source.

Single issues and back volumes: Information can be obtained through the Publisher.

Advertising: Correspondence and rate requests should be addressed to the Publisher.

Bibliographic indices: This journal is regularly listed in bibliographic services, including *Current Contents®*.

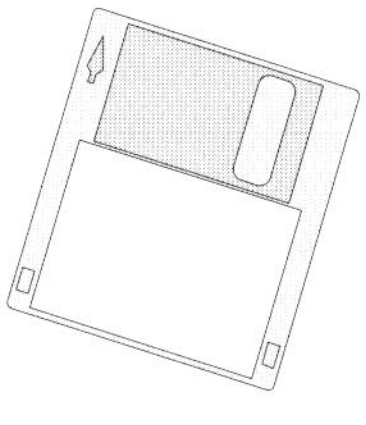

Disk Submission

A **comprehensive version** of these guidelines for the **submission of manuscripts and illustrations on disk** can be accessed online under **www.karger.com/journals/disksub.htm** or requested as a leaflet from the editorial department.

Please follow the general rules on style and arrangement in the **'Guidelines for Authors'** of the journal. Submit a **double-spaced printout** of your manuscript matching the file on disk exactly.

Technical Information

Store only data pertinent to this manuscript. Include your fonts (screen and printer fonts for Macintosh). For a **text only** manuscript, use a 3.5-inch disk. For **text and illustrations,** use 3.5 or 5.25 Magneto Optical Disc; CD-ROM (DOS or hybrid); iomage Zip/DOS (100 MB), or 44 MB, 88 MB or 200 MB SyQuest. Compress extensive data using ZIP format.

Data Description

Label your disk with the **title of the journal or book** for which it is submitted; a **short version** of the paper's title; your **name,** postal and e-mail address, fax and phone number; **hard- and software** used, including the version; and the **complete list of the file names.** File names consist of eight characters before the dot identifying the file and three characters after the dot identifying the application.

Text and Format

The **preferred word-processing package is MS-Word.** Other commonly used PC text programs and ASCII formats are accepted.
Update your data when you revise subsequent to the reviewer's or editor's decision and mark revisions clearly. Manuscripts accepted for publication are subject to **copy editing:** changes and/or questions will be marked in the manuscript, which will be returned to you with your disk and the galley proofs for checking.
Enter your text continuously flush left. Do not use any layout functions. Do not split words at the end of a line. **Do not use indentations.** Use a blank line before and a hard return at the end of paragraphs.
Arrange headings flush left. Insert only **one space after punctuation marks.** Insert Greek letters, mathematical symbols etc. with your word-processing program. Use **boldface and italics** as well as sub- and superscript where appropriate.
Tables are part of the text and should be placed at the end of the text file. Separate the individual parts of a table by lines. Use the tabulator instead of spaces. Format operation data on one line in the same column. **Legends to illustrations are part of the text** and should also be listed at the end of the text file.

Illustrations

Store illustrations as separate files: do not integrate them in your text. Export **line drawings/vector graphics** as TIF, EPS or WMF format. Use Photoshop for processing and retouching **scanned halftone images.** Save the original scan and the processed version. Export **black and white or color images** as TIF or EPS format in their anticipated size in print.
For detailed information on how to submit illustration data on disk please consult the leaflet or **www.karger.com/journals/disksub.htm.**

KARGER

Fax+41 61 306 12 34
E-Mail karger@karger.ch
www.karger.com

The Journal Home Page is available at:
www.karger.com/journals/cto

KI 98 DISK

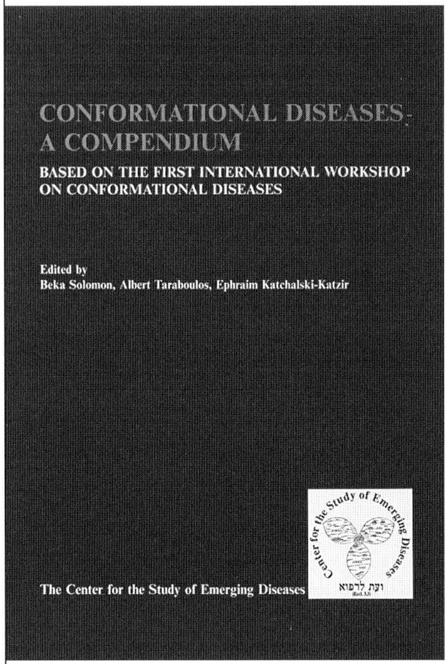

Conformational Diseases – A Compendium

Editors
Beka Solomon
Albert Taraboulos
Ephraim Katchalski-Katzir

Molecular Biology; Neurodegenerative Diseases, Protein Conformation, Pathology, Genetics, Biochemistry

Selected Contents

'Conformational diseases' are a group of heterologous disorders in which a constituent host protein undergoes a change in conformation or fluctuation in shape with resultant self-association and deposition. The etiology of many diseases has now been traced to proteins that are either misfolded from the start or have undergone post-translational conformational alterations. The articles in this compendium reflect the recent advances and exponential growth of the field. They also provoke many questions. How do malfolded proteins cause disease? How are such diseases transmitted from one individual to another, or from animals to humans? And how can one prevent these conformational alterations or combat their deleterious effects? These and many other questions remain to be answered before we can achieve a clearer picture of conformational disorders.
The beautifully produced compilation is divided into seven sections: Clinical aspects, theoretical considerations, experimental studies, models *in vivo* and *in vitro*, genetics, transgenetics and cell biology, prospects for treatment, and further considerations. Contributions include original articles by leading scientists, including Nobel laureate G. Carleton Gajdusek, Charles Weissmann, Shuguang Zhang, Michael Altman and Alexander Rich, Michel Goldberg, F.J. Stevens, M. Schiffer and Y. Argon. These are complemented by seminal papers that have already appeared in top-ranking journals, written by outstanding scholars such as Nobel laureate Stanley Prusiner, philosopher Henri Atlan, Christopher Dobson, Reed Wickner, Jean Sipe and many others. The resulting compilation is a comprehensive handbook on conformational diseases that encompasses many of the central elements in the field.

For the complete contents please see: www.karger.com/confo

KI 02251

Conformational Diseases – A Compendium
Based on the First International Workshop on Conformational Diseases
Editors: Solomon, B.; Taraboulos, A.; Katchalski-Katzir, E.
XIV + 324 p., 155 fig., 58 in color, 43 tab., soft cover, 2001
CHF 290.– / EUR 207.– / USD 252.25
Prices subject to change
EUR price for Germany, USD price for USA only
ISBN 3–8055–7410–X

Distributed in Israel exclusively by the Center for the Study of Emerging Diseases

KARGER

Order Form

Please send: ____ copy/ies

Postage and handling free with prepayment

Payment:
☐ Check enclosed ☐ Please bill me

Please charge to my credit card
☐ American Express ☐ Diners ☐ Eurocard
☐ MasterCard ☐ Visa

Card No.: ____________

Exp. date: ____________

Orders may be placed with any bookshop, subscription agency, directly with the publisher or through a Karger distributor.

Fax: +41 61 306 12 34
S. Karger AG, P.O. Box, CH–4009 Basel (Switzerland)
E-Mail karger@karger.ch, **www.karger.com**

Name/Address:

Date:

Signature:

Cells Tissues Organs

in vivo, in vitro

Founded 1945 as 'Acta Anatomica' by R. Chambers, G. Glimstedt, T. Peterfi and G. Wolf-Heidegger
Continued 1962–1974 by E.A. Boyden, 1955–1980 by A. Delmas, 1972–1980 by F. Walberg, 1945–1980 by G. Wolf-Heidegger, 1981–1988 by R. O'Rahilly, Davis, Calif., 1989–1990 by G.E. Goslow, Jr., Providence, R.I., 1981–1992 by W. Lierse, Hamburg

KARGER

Printed in Switzerland on acid-free paper by Reinhardt Druck, Basel

Appears monthly: 3 volumes per year (12 issues)

Guidelines for Authors

Cells
Tissues
Organs

Aims and Scope

Cells Tissues Organs is an international forum for communication among scientists interested in current research on functional morphology and morphogenesis at all levels of organization, from subcellular to macroscopy, with emphasis on humans and higher vertebrates. The journal offers rapid exchange of novel and significant information gained from experimental and theoretical investigations on all aspects that are pertinent to the formation and transformation of biological structure.

Topics of special interest include

Developmental and reproductive biology, cell biology, tissue engineering, artificial organs, in vitro systems and transplantation. Papers highlighting the links between structural and functional aspects are preferred.

Types of Papers

The journal publishes: (1) Original papers on experimental research. (2) Theoretical contributions. These will appear in the section 'Hypotheses and Concepts' and may include provocative hypotheses, provided that they are innovative, well founded and show promise for new avenues of research. (3) Short reviews/perspectives. While (1) and (2) are submitted contributions, reviews/perspectives (3) are invited. Suggestions concerning reviews on current topics of high interest are welcome.
All manuscripts must be accompanied by a cover letter signed by all authors. Assurance should be given that the manuscript is not under simultaneous consideration by any other publication. Reported studies using animal and human subjects must have been approved by the appropriate institutional review boards.

Submission

Manuscripts from North and South America should be sent to:

Prof. A.W. English, PhD
Emory University School of Medicine
Dept. of Anatomy and Cell Biology
Atlanta, GA 30322 (USA)
Fax +1 404 727–3677
E-Mail cto@cellbio.emory.edu

Manuscripts from Japan and South-East Asia to:

Prof. Dr. Osamu Ohtani
Department of Anatomy II
Toyama Medical and Pharmaceutical University
2630 Sugitani
Toyama, 930-01 (Japan)
Fax +81 764 34 5010

Manuscripts from all other countries to:

Prof. H.-W. Denker
S. Karger AG
Editorial Office 'Cells Tissues Organs'
PO Box
CH–4009 Basel (Switzerland)
Tel. +41 61 306 13 57
Fax +41 61 306 14 34
E-Mail l. haas@karger.ch

Mailing address for courier deliveries only:

Allschwilerstr. 10
CH–4055 Basel (Switzerland)

All manuscripts are subject to peer review. Submission of an article for publication implies the transfer of the copyright from the author to the publisher upon acceptance. Accepted papers become the permanent property of Cells Tissues Organs and may not be reproduced by any means, in whole or in part, without the written consent of the publisher. It is the author's responsibility to obtain permission to reproduce illustrations, tables, etc. from other publications.

Format and Arrangement

Manuscripts should be written in English and be as concise as possible. They should be submitted in quadruplicate (with four sets of original illustrations), typewritten double-spaced on one side of the paper, with a wide margin. *Lines should be numbered in the margin.*

Title page: Title, the authors' names, and the institute where the work was conducted, a short title for use as running head, the full address of the author to whom correspondence should be sent. Please also supply phone and fax numbers, as well as your e-mail address. If different, indicate in addition a full address for revisions and proof reading separately.

Key words: For indexing purposes provide a list of up to 5 key words.

Abstract: Each paper needs an abstract of up to about 20 lines summarizing the essential new information communicated.

Subheadings: Introduction; Materials and Methods; Results; Discussion.

Small type: Paragraphs which can or must be set in smaller type (methods, etc.) should be indicated with a 'p' (petit) in the margin on the left-hand side.

List of Abbreviations: Provide a list of the abbreviations used in your paper. Please list them on a separate page of the manuscript. The list will be printed as a footnote at the bottom of the first page of your article.

Footnotes: With the exception of the list of abbreviations, footnotes should be avoided. When essential, they are numbered consecutively and typed at the foot of the appropriate page.

Nomenclature: The nomenclature used should conform, wherever possible, to the current edition of the Nomina Anatomica (Terminologia Anatomica), Nomina Anatomica Veterinaria, or to the Nomina Anatomica Avium.

Tables and illustrations: Tables and illustrations (both numbered in Arabic numerals) should be prepared on separate sheets. Tables require a heading and figures a legend, also prepared on a separate sheet. For the reproduction of illustrations, only good drawings and original photographs can be accepted; negatives or photocopies cannot be used. Due to technical reasons, figures with a screen background should not be submitted. When possible, group several illustrations on one block for reproduction (max. size 181 × 223 mm) or provide crop marks. On the back of each illustration, indicate its number, the author's name, and 'top' with a soft pencil.

Color illustrations: During the review process, the decision will be made as to which color illustrations of a manuscript can be reproduced free of charge. Authors will be informed of this at the time of acceptance of their paper. Otherwise, these illustrations will be reproduced at the author's expense. Up to 6 color illustrations per page can be integrated into the text at the reduced price of CHF 660.– / USD 545.00 per page.

References: References in the text should be made by author's name followed by the year of publication, arranged chronologically, then alphabetically. When there are more than two authors, use the first author's name and et al. When references are made to more than one paper by the same author, published in the same year, designate them as a, b, c, etc. The list of references should include only those publications which are cited in the text. Material submitted for publication but not yet accepted may be cited in the text as 'unpublished data' but must not be included in the reference list. In the list of references, arrange references alphabetically listing all authors, then year of publication. Abbreviate journal names according to Index Medicus.

Examples

(a) Papers published in periodicals: Sun, J., H. Koto, K.F. Chung (1997) Interaction of ozone and allergen challenges on bronchial responsiveness and inflammation in sensitised guinea pigs. Int Arch Allergy Immunol *112:* 191–195.

Hadler, W.A., S.R. Silveira (1981) Técnica histoquímica que permite discriminar o colesterol de substâncias afins e de seus ésteres. Cienc Cult *33(suppl):* 586–587.

(b) Monographs: Matthews, D.E., V.T. Farewell (1996) Using and Understanding Medical Statistics, ed 3, revised. Basel Karger.

(c) Edited books: Parren, P.W.H.I., D.R. Burton (1997) Antibodies against HIV-1 from phage display libraries: Mapping of an immune response and progress towards antiviral immunotherapy; in Capra, J.D. (ed): Antibody Engineering. Chem Immunol. Basel, Karger, vol 65, pp 18–56.

Further details: Authors may consult the 'Guide to Writing Articles in English' by R. O'Rahilly (Acta Anat *131*: 1–2, 1988).

Galley proofs

Unless indicated otherwise, galley proofs are sent to the first-named author and should be returned with the least possible delay. Alterations made in proofs, other than the correction of printer's errors, are charged to the author.

Reprints

Order forms and a price list are sent with the proofs. Orders submitted after the issue is printed are subject to considerably higher prices.

KARGER

Fax+ 41 61 306 12 34
E-Mail karger@karger.ch
www.karger.com

The Guidelines for Authors are available at:
www.karger.com/journals/cto/cto_gl.htm

Cells
Tissues
Organs

Somatic Stem Cells and Their Plasticity

Guest Editors
Albrecht M. Müller, Würzburg
Kelly M. McNagny, Vancouver

25 figures, 18 in color, 5 tables, 2002

KARGER
Basel · Freiburg · Paris · London · New York ·
New Delhi · Bangkok · Singapore · Tokyo · Sydney

S. Karger
Medical and Scientific Publishers
Basel · Freiburg · Paris · London
New York · New Delhi · Bangkok
Singapore · Tokyo · Sydney

Drug Dosage
The authors and the publisher have exerted every effort to ensure that drug selection and dosage set forth in this text are in accord with current recommendations and practice at the time of publication. However, in view of ongoing research, changes in government regulations, and the constant flow of information relating to drug therapy and drug reactions, the reader is urged to check the package insert for each drug for any change in indications and dosage and for added warnings and precautions. This is particularly important when the recommended agent is a new and/or infrequently employed drug.

P.O. Box, CH–4009 Basel (Switzerland)
Printed in Switzerland on acid-free paper by
Reinhardt Druck, Basel
ISBN 3–8055–7406–1

KARGER

Fax + 41 61 306 12 34
E-Mail karger@karger.ch
www.karger.com

Vol. 171, No. 1, 2002

Cells
Tissues
Organs

Contents

KARGER

Fax + 41 61 306 12 34
E-Mail karger@karger.ch
www.karger.com

Access to full text and tables of contents,
including tentative ones for forthcoming issues:
www.karger.com/journals/cto/cto_bk.htm

Cells
Tissues
Organs

Cells Tissues Organs 2002;171:5

Editorial Note

With this edition on 'Somatic Stem Cells and Their Plasticity' CTO continues to publish what is intended to become a series of special issues devoted to recent developments in stem cell research, started in 1999 with 'Embryonic Stem Cells as a Developmental Model in vitro' (Vol. 165/3–4; A.M. Wobus and K.R. Boheler, eds.), a publication that was very well received by the scientific community. We expect that the present issue may find at least as much attention as the previous one: Recently an exploding number of reports has demonstrated an astonishing plasticity of somatic stem cells, findings that were totally unexpected at earlier times and that may revolutionize our views of embryonic development as well as our prospects of possible therapeutic uses of somatic vs. embryonic stem cells. I am very happy that Albrecht Müller and Kelly McNagny have accepted to take care of this publication project and have invested the enormous amount of time and effort that was needed to bring this to fruition. It is due to their enthusiasm that it was possible to bring together an excellent spectrum of contributions from various leading laboratories.

For CTO this special issue intends to underline and deepen our new focus on the emerging fields of transplantation biology, tissue engineering, artificial organs and in vitro systems and our devotion of basing research in these fields on modern concepts of cell and developmental biology. CTO plans to continue on in this direction.

Hans-Werner Denker

KARGER
Fax + 41 61 306 12 34
E-Mail karger@karger.ch
www.karger.com

1422–6405/02/1711–0005$18.50/0

Accessible online at:
www.karger.com/journals/cto

Cells
Tissues
Organs

Cells Tissues Organs 2002;171:6–7

Preface

Embryogenesis of higher organisms results in the generation of an adult and fertile organism from a single cell, the zygote. This is achieved through the judicious use of 'stem cells' though not all stem cells of the embryo persist into adulthood. Adult somatic stem cells arise during embryogenesis and have the unique properties of being able, at the clonal level, to self-maintain their own numbers and to differentiate into mature progeny of a particular tissue. Thus, tissue homeostasis is maintained despite rapid cell turnover by the continuous contribution of tissue stem cells. Classic examples of such cells in higher vertebrates include stem cells in the bone marrow, gut and skin. To the scientists these tissue stem cells offer a precious source of undifferentiated material for the study of cell maturation and specialization. To the clinician, these cells represent a potential source of material for the cure and replacement of damaged organs and tissues. The promise they hold for future therapy is enormous. Any disease resulting from genetically or environmentally damaged tissues could, in principal, be repaired through a stategic use of stem cells. Yet this is critically dependent on a basic understanding of stem cell biology and function.

Through years of study a number of central concepts and dogmata have emerged to explain the behavior of stem cells. As with all dogmata in science, these models are only as valid and robust as our ability to test them. Some properties of somatic stem cells that were assumed, until recently, to be indisputable include the following: (1) Differentiation is a one-way street i.e. re-acquisition of a more primitive phenotype is forbidden. (2) Trans-differentiation is also forbidden i.e. once a cell has acquired a differentiated phenotype, it loses the ability to adopt an alternative fate. (3) Stem cells that give rise to a particular tissue are 'fixed' in terms of their potentiality and are unable to contribute to the formation of other tissues.

Challenges to the first two concepts have been the topic of extensive debate for a number of years. Initial reports focused on examples of 'de-differentiation' and 'trans-differentiation' in tumors and cell lines. In many cases, these phenotypic transformations were generally attributed to abnormal 'phenomenology' associated with oncogenesis. More recently, however, it has become clear that, given the appropriate combination of transcription factors and extracellular cues, phenotypic shifts can be induced in ways that would have been unthinkable just a few years ago.

More current and even more controversial are recent reports that tissue-specific stem cells, which had previously been thought to be committed to the formation of a given tissue, show a high degree of plasticity and are capable of generating other tissues when provided with the appropriate microenvironment. Until recently, the idea that a mesodermally derived stem cell would be able to give rise to neuro-ectodermal tissues was considered pure, unadulterated heresy. And yet now we are confronted with a spate of papers that challenge this very concept. Recent reports of bone marrow-derived cells giving rise to muscle, brain, liver and skin, and neural stem cells and stem cells in the muscle generating blood are challenging earlier concepts of stem cell boundaries and cell fate. However, one crucial and technically challenging experiment is required to make these current experiments conclusive: it is critical to show that these phenomena occur at the clonal level i.e. that the very same cell that gives rise to blood can also give rise to other tissues and vice versa. With this important caveat aside, several key questions remain: Are the current fate maps of tissue-specific stem

KARGER
Fax + 41 61 306 12 34
E-Mail karger@karger.ch
www.karger.com

1422–6405/02/1711–0006$18.50/0

Accessible online at:
www.karger.com/journals/cto

cells wrong and can they cross barriers that we previously considered inviolate? Is the plasticity of somatic stem cells a relic of our evolution? And finally, is plasticity of somatic stem cells of physiological relevance or merely an experimental phenomenon?

This issue of *Cells Tissues Organs* contains a timely series of reviews that serve as a primer for basic concepts surrounding stem cell biology. They explore a broad range of topics including a basic definition of 'stem cell-ness' (Loeffler and Roeder), the genetic control of stem cell longevity (de Haan), transcriptional regulation of stem cell fate (Bruno et al.; Nielsen et al.) the brewing debate on tissue stem cell plasticity (Kirchhof et al.; MacKenzie and Flake; Vescovi et al.). The goal of this issue is to provide a conceptual scaffold of our current knowledge of stem cell biology and to highlight areas of current uncertainty and direct interest towards future experimentation.

Regardless of whether the current dogma holds or falls, the recent debate has helped marshal interest and resources into the study of this exciting field. This will undoubtedly provide us with new and exciting findings about the capabilities of stem cells and maybe new dogma for future challenges.

Kelly M. McNagny, Vancouver
Albrecht M. Müller, Würzburg

Cells
Tissues
Organs

Cells Tissues Organs 2002;171:8–26

Tissue Stem Cells: Definition, Plasticity, Heterogeneity, Self-Organization and Models – A Conceptual Approach

Markus Loeffler   Ingo Roeder

Institute for Medical Informatics, Statistics and Epidemiology, University of Leipzig, Leipzig, Germany

Key Words
Tissue stem cells · Definition · Dynamic model · Plasticity · Self organization · Simulation

Abstract
The classical definition of adult tissue stem cells (TSC) is fundamentally based on a functional perspective. A TSC is an undifferentiated cell, capable of proliferation, self-renewal, production of a large number of differentiated functional progeny, regenerating tissue after injury and a flexibility in the use of these options. Here, we discuss the necessity for amending this definition in the light of recent insight into stem cell biology regarding stem cell heterogeneity, lineage plasticity, clonal fluctuation and cell–environment interactions. We conclude that the definition needs amendments. A decade ago the flexibility criterion has attracted little attention but recent findings have indicated its importance. Flexibility and reversibility of tissue and lineage specification *(tissue plasticity)* and of properties within a tissue *(within-tissue plasticity)* have major implications with regard to concepts of stem cell function. We advocate to give up the view of TSC as being entities with a preprogrammed development and to replace it by a concept that makes the capabilities for flexible and regulated tissue self-organization based on cell–cell and cell–environment interactions the new paradigm. This concept would permit to incorporate the context-dependent lineage plasticity, within-lineage plasticity and generation of stem cell heterogeneity as a result of a dynamically regulated process. Such concepts need a rigorous examination by formal modeling including simulation studies. We provide some general ideas on how to proceed with such theories and illustrate this with worked models for tissue stem cells of the hematopoietic system and the intestinal epithelium.

Introduction

We consider some of the problems involved in the definition and understanding of stem cell systems in mammalian tissues. Over 10 years ago we have proposed a definition of tissue stem cells (TSC) which has been widely used and cited [Potten and Loeffler, 1990]. The basic concept of a functional definition has proven valid. An ingredient of this definition, not always appreciated, was that it does not require stemness as an explicit attribute of cells but rather considers it as a functional endpoint. We review this 'classical' definition in the light of novel findings e.g. on plasticity, stem cell heterogeneity and cell–growth environment interactions and propose an amended and annotated definition. We also discuss the implications

KARGER
Fax + 41 61 306 12 34
E-Mail karger@karger.ch
www.karger.com

1422–6405/02/1711–0008$18.50/0

Accessible online at:
www.karger.com/journals/cto

Markus Loeffler
Institute for Medical Informatics, Statistics and Epidemiology
University of Leipzig, Liebigstr. 27
D–04103 Leipzig (Germany)
Tel. +49 341 9716100, Fax +49 341 9716109, E-Mail Loeffler@imise.uni-leipzig.de

with regard to basic concepts on functioning of tissue stem cell systems and attempts to characterize a molecular profile of TSC. We further argue that some of the conceptual problems involved in understanding TSC could be overcome if more comprehensive and predictive theories of stem cell systems were available. We provide a list of criteria such theories should fulfill. For illustration we report on models of the hematopoietic system and for the intestinal epithelium. Both models are fully consistent with the amended definition but do not imply explicit assumptions on stemness properties in the cells. They rather indicate that a concept of controlled within-tissue plasticity utilizing a spectrum of potentials might be a basic component in such theories.

Several ideas and problems presented subsequently have been discussed on the Workshop 'Tissue Stem Cells – Models and Concepts' held in Leipzig on June 12–13th, 2001 preceding the 24th Meeting of the European Study Group for Cell Proliferation. The authors thank the participants (see acknowledgement) for their contributions and suggestions.

Classical Definition of Tissue Stem Cells

We start by recapitulating our previous definition for tissue stem cells [Potten and Loeffler, 1990; Loeffler and Potten, 1997] in order to provide the basis for an amended and annotated version. For a detailed discussion of important notions such as differentiation, maturation, self-maintenance, self-renewal, or regeneration we refer to the already above cited articles.

The definition for tissue stem cells proposed by Potten and Loeffler, 1990 was:
Stem cells of a particular tissue are

- undifferentiated cells (relative to a functional tissue) capable of
- proliferation;
- production of a large number of differentiated functional progeny;
- self-maintenance of their population;
- regeneration of the tissue after injury;
- flexibility in the use of these options.

Comments on This Definition of Tissue Stem Cells

There are several key features in this definition.

Firstly, stem cells are defined by virtue of their functional attributes and not by an explicit directly observable characteristic. The choice of a *functional definition* is inherently consistent with the biological role of a stem cell particularly linked to the functional tissue regeneration feature. This kind of definition, however, imposes difficulties since in order to identify whether or not a cell is a stem cell, its function has to be tested. This inevitably demands that the cell must be manipulated experimentally, which may actually alter its properties. We will return to this circular problem later.

Secondly, tissue stem cells are assumed to be undifferentiated compared with the functional end cells of the particular tissue to which they give rise. This definition is essentially a relative one as it relates the stem cells to the functional end cells, or to cells at earlier stages of the development. Apparently, this definition is compatible with the existence of various stem cells of different tissues as well as of a hierarchy of stem cells for one particular tissue. It is possible that there are specific differentiation markers which would enable a distinction of stem cells in relation to each other and in relation to the functional cells they are eventually producing. The *relativity of stemness* is an essential feature to keep in mind and one has to be specific with respect to the particular experimental circumstances.

Thirdly, cells originating from stem cells may loose some of the properties. This gives a possible way to classify descendent transit and mature cells (see below).

Fourthly, not all of these criteria have the same weighting. The most important criteria certainly is the regeneration of the tissue after injury in conjunction with the self-maintenance criterion. The latter, however, needs some more careful consideration as will be discussed below.

Fifthly, there was no explicit assumption made in the definition on the kind of mechanisms involved in creating stem cell like behavior. In particular no mentioning was made, on the coupling of proliferation with differentiation (e.g. symmetric or asymmetric cell division).

Sixthly, an essential aspect of the definition is the flexibility criterion. It was primarily introduced to account for potential regulatory modulations in adjusting the functional operation of stem cells to some demands (e.g. injury or challenges) in order to assure tissue function. Furthermore, there were reports on varying estimates of stem cell numbers in the same tissue depending on the measurement process. Introducing flexibility as a key aspect had the conceptual consequence that we include the definition of stem cells in different functional states. Cells or populations of cells actually fulfilling all these criteria at a given instance are called *actual stem cells*, while those not actually expressing these capabilities at a particular moment in time, although they possess these capabilities, will be termed *potential stem cells*. It may be possible for a stem cell to cease proliferation, i.e. become *quiescent*, in which case it does not act as an actual stem cell, but since it can re-enter the cycle it has the potential to act as a stem cell. Likewise a transit cell (see below) may not normally self-maintain, but may do so under special circumstances, thereby representing a potential stem cell. We choose the word actual in preference to the term functional, which has been used previously [Cairnie et al., 1965; Steel, 1977; Wright and Alison, 1984].

Finally, in the stem cell definition given above, *pluripotency* was not requested as a prerequisite of stemness. The ability to produce progeny that differentiate down various lineages (pluripotency) is not necessarily a property of tissue stem cells per se, although it appears that stem cells in many tissues possess this capability as a feature of regeneration of the tissue. However, we do not want to exclude the possibility that a stem cell system may only produce one type of cells.

Amended and Annotated Revised Definition

The above mentioned classical definition has been successful in many respects as was extensively discussed in the workshop on 'Tissue Stem Cells' in Leipzig in 2001. The participants agreed that the following features were essential and should be included in any revision of the definition:

- The functional definition of tissue stem cells represents the gold standard. This is considered most appropriate to reflect the biological nature. At present this definition cannot be replaced by definitions based on surrogate criteria e.g. on phenotypic markers, gene expression profiles or other molecular definition. All these definitions would miss the essential and complex functional aspects.
- The flexibility criterion is an essential one and it should be given a more prominent role in the light of recent experimental findings.
- A change in perspective from a cellular view to a tissue view is important and should be incorporated into the definition of tissue stem cells. Rather than to focus on cells alone it is important to consider the entire self-organizing system of cells and growth environment comprehensively.

In addition, over the past decade experimental evidence has accumulated that makes amending and sharpening the definition desirable in several further respects. The main aspects we like to include subsequently relate to the following topics:

- consideration of the cell – growth environment interaction in order to take the system character into account;
- extension of the flexibility criterion to incorporate lineage and within-tissue plasticity;
- consideration of stem cell heterogeneity and the implications for the self-maintenance criterion.

We suggest to amend the above list of criteria as follows:

Amended Definition of Tissue Stem Cells

Stem cells of a particular tissue are:

(S1) a potentially heterogeneous population of functionally undifferentiated cells, capable of:
(S2) homing to an appropriate growth environment;
(S3) proliferation;
(S4) production of a large number of differentiated progeny;
(S5) self-renewing or self-maintaining their population;
(S6) regenerating the functional tissue after injury with
(S7) flexibility and reversibility in the use of these options.

Subsequently we discuss these amendments one by one:

(S1) *'Stem cells of a particular tissue are a potentially heterogeneous population of functionally undifferentiated cells ...'*

There are two closely linked modifications in this phrasing compared with the classical definition. We suggest to relate the definition to a population of cells rather than to individual cells for several reasons. Any determination of self-maintenance or self-renewal (see below) requires measurements on populations of cells over a certain period of time. The criterion of *functionally undifferentiated cells* practically implies that one negatively selects cells on a limited set of differentiation attributes (e.g. phenotypic or genotypic markers). Hence, the suspected stem cells under examination are always a population of remainder cells which may be heterogeneous with regard to the features not used for the selection process. The tissue regeneration criterion also requires the population view. Even if the tissue is regenerated from a single cell it implies regeneration of a population of stem cells. A population view is furthermore required to account for the flexibility and reversibility criterion. It is basically linked to the phenomenon of a cellular and functional heterogeneity which is the second addition to the criterion.

As an example for a heterogeneous population we can refer to the hematopoietic stem cells [Uchida et al., 1993; Lord, 1997]. Phenotypic heterogeneity has been described with regard to various markers (e.g. CD34, CD38, c-kit, Sca1). They can be used in selection procedures but are probably only indirectly linked to stem cell functioning. While it has previously been thought that CD34 positivity selects stem cells it has become clear recently that also CD34-negative cells can be effective stem cells and that the cells can even alter this CD34 property [Sato et al., 1999].

(S2)–(S6) *'... capable of ...'*

We maintain and even extend the concept of capabilities. We consider stemness as an option to behave in one or more ways in the future rather than a property realized actually. This implies a prospective rather than a momentary view as one has to challenge the cells in particular ways to determine the capabilities. The following capabilities, (S2)–(S6), are independent of each other.

(S2) *'... homing to an appropriate growth environment.'*

We introduce this criterion as a new one to reinforce the aspect of a system which is not only driven by characteristics fixed in the cells but requires a specific and adequate interaction of the cells with a growth environment. The growth environment can have different functional and spatial architectures depending on the tissue. In the intestinal jejunal or colonic crypt e.g. stem cells are located at the bottom of the crypt from where the cell flow starts [Wright and Alison, 1984; Potten, 1998]. Apparently this location is a spatial niche difficult for other cells to occupy. For the hematopoietic system functional/spatial niches have been suggested already many years ago [Trentin, 1971; Schofield, 1978, Lord and Wright, 1984]. The picture may, however, be more complicated as recent insight becomes available on the non-hematopoietic stromal cells involved [Wineman et al., 1996; Gordon et al., 1997; Lemischka, 1997; Oostendorp and Dormer, 1997], about the interactions mediated by adhesion molecules [Vermeulen et al., 1998; Roy and Verfaillie, 1999; Kapur et al., 2001], about the role of paracrine growth factor stimulation [Bodine et al., 1991; Neben et al., 1994; de Haan et al., 1995], about gap junctions between stem cells [Rosendaal et al., 1997; Cancelas et al., 2000], and pseudopod connections [Francis et al., 1998; Frimberger et al., 2001]. The difficulties to produce fully functional hematopoietic stem cells in culture is another hint to this complexity. This indicates that in fact hematopoiesis is based on a complex functional and spatial arrangement of growth environments which may be less of a spatial but more of a functional architecture. E.g. in murine hematopoiesis both the bone marrow and the spleen can support granulocytopoiesis and erythropoiesis. However, there is a clear bias towards erythropoiesis in the spleen [Wichmann and Loeffler, 1985].

The homing criterion is introduced to request some sign of a real interaction between the cells and the growth environment. In some stem cell systems we even know about the possibility of an active migration into the growth environment. In the bone marrow the present view is now that there is rather intensive migratory traffic with cells actively searching for adequate environments [Blau et al., 2001; Frimberger et al., 2001]. Similar findings hold for the epidermis where stem cells are able to migrate from the hair follicles to the basal epidermal layers [Miller et al., 1997]. Migration, however, is not deemed a universal feature. E.g. it seems very unlikely that tissue stem cells can migrate from one crypt to another in the intestinal system.

(S3) *'... proliferation.'*

Proliferation is apparently an active feature in actual stem cells. The criterion is redundant if tissue regeneration or clonal expansion of an active clone is considered. We prefer to maintain this capability in the definition, because it explicitly permits the status of silent tissue stem cells (e.g. in G_0) from which they can be activated again [Fleming et al., 1993; Cheshier et al., 1999; Goodell, 1999; Tajima et al., 2000].

(S4) *'... production of a large number of differentiated progeny.'*

This criterion places tissue stem cells at the root of a cellular amplification process connected with differentiation and production of functional end cells. Until recently this capability was assessed mostly by the use of colony formation studies in vivo or in vitro. More rigorous ways have become available by possibilities to examine clonal development in vivo in greater detail. Genetic marker studies based on chimerism of competing cell populations or based on the introduction of unique markers into single cells (e.g. random mutation or retroviral marking) permit to assay the clonal development. However, not all tissue stem cells may actually produce such a clone at the time of examination. Furthermore, there is evidence that the clonal composition of a tissue evolves with fluctuations in time and space. This has been observed in hematopoiesis and in epithelia [Winton et al., 1988; Jordan and Lemischka, 1990; Abkowitz et al., 1996; Bjerknes and Cheng, 1999].

(S5) *'... self-renewing or self-maintaining their population.'*

The classical definition of TSC requested the capability of self-maintenance as an indispensable feature. Please note, self-maintenance implies only 'keeping at an existing state' and it is often used to describe particularly the numerical maintenance. In homogenous stem cell populations maintaining the population size is obviously identical with maintaining the qualitative aspects of it. As we extend our definition to also encompass heterogeneous populations of tissue stem cells (criterion (S1)) numerical maintenance is not a sufficient requirement. A heterogeneity in TSC composition may imply that some subpopulations behave more effective with regard to some assays than others. Hence, it is possible to have a bias in the composition of TSC that is not revealed by the numerical criterion of maintaining the entire population. To sharpen the definition in this regard we propose to amend the criterion by the more rigid requirement of self-renewal. Self-renewal implies that any tissue stem cell taken from a heterogeneous population should be capable of generating a heterogeneous population of TSC with the same (or similar) number *and* composition as the initial population from which it originated.

Self-renewal in this sense is a population property again. This criterion is apparently closely linked to reversibility of some cellular developments.

(S6) *'... regenerating the functional tissue after injury.'*

This criterion describes the most important endpoint of tissue stem cell functioning. The cells must be capable of giving rise to cell populations which can generate a functional tissue. In the latter respect we have made the criterion more rigorous in demanding the functional composition and competence of functional cells. As has been pointed out by others [e.g. Anderson et al., 2001] it is not sufficient to show that a tissue generated from some presumptive tissue stem cells looks like a functional tissue but that the clonally derived cells in fact satisfy all functional criteria. This aspect of regenerating functional heterogeneity also implies in our mind the possibility of lineage specification and encompasses the pluripotency aspect of a functional tissue. Hence, pluripotency is not considered as an independent but as a surrogate marker for functional capability related to criterion (S6). The criterion of functional tissue regeneration also implies the features of self-organization, homeostasis and of longevity. Long-term tissue regeneration from a small pool of tissue stem cells also implies some kind of self-renewal, dynamically linking the criteria (S4) and (S5) in the long run.

(S7) *'... with flexibility and reversibility in the use of these options.'*

The flexibility criterion was already included in the classical definition as a conceptual implication and prediction. Little direct experimental evidence was available at that time. We now see emerging and fascinating evidence that flexibility and in addition reversibility are fundamental properties of tissue stem cells. In terms of this definition flexibility and reversibility are related to the criteria (S2)–(S6). This implies the possibility that tissue stem cells can leave and re-enter proliferation, that they can alter the degree of self-renewal in either direction, that the degree of clonal expansion and the functional competence of a regenerated tissue may vary. These variations may affect functional competence or lineage specification in a tissue with several cell lineages. Reversibility has been added to this criterion. Again, reversibility is requested as a *potential* property which may not play a role in an actual biological situation. Whether flexibility and reversibility are actually utilized depends on the status of the tissue and on the regulatory processes that are involved. We speculate that this latter criterion plays an important role in tissue evolution. Tissues must be designed in such a way that they can respond to a large variety of mechanical and chemical insults and to different metabolic circumstances. Hence, a capability to respond with some degree of flexibility, with regard to timing, cell production, cell functioning and cell–cell interaction should have been an important selection advantage. We therefore believe that it is one of the essential properties of tissue stem cells to be able to react to such a variety of demands.

There is an accumulating experimental evidence for flexibility and reversibility. We like to highlight a few of these preferably related to the hematopoietic system. Many authors have described the variability in the proliferative status of hematopoietic stem cells. Perhaps the most interesting findings are that some cells may leave the cell cycle for many days and even months, but that they re-enter almost all at some time as was shown by several long-term labeling studies [Bradford et al., 1997; Cheshier et al., 1999]. Quesenberry and colleagues provide evidence for reversible changes of the stem cell phenotypes involving differentiation profiles, adhesion protein expression and engraftment/homing behavior associated with the cell cycle status or the point in the circadian rhythm [Habibian et al., 1998; Quesenberry et al., 1998]. There is increasing evidence that the expression of some surface markers like CD34 on hematopoietic stem cells is not constant but may fluctuate. The property can be gained and lost without affecting the stem cell quality [Goodell, 1999, Sato et al., 1999]. Mueller and his group investigated globin switching of hematopoietic stem cells in the blastocyste growth environment. They showed that the switch from embryonic/fetal-type to adult-type globin is reversible [Geiger et al., 1998]. Furthermore there is a lot of indirect evidence for fluctuations in the stem cell population based on the clonal composition of functional cells. Chimerism induced by transplantation maneuvers in cats and mice has been shown to fluctuate with time [Van Zant et al., 1992; Abkowitz et al., 1996; Abkowitz et al., 2000] indicating variations in the composition of active and inactive tissue stem cells. Similar observations were made following retroviral marking [Jordan and Lemischka, 1990; Van Zant et al., 1991; Drize et al., 1997]. For the intestinal crypt there is good evidence for a competition process of tissue stem cells within the individual crypts. This competition leads to a fluctuation of the clonal composition with a dynamic instability [Winton and Ponder, 1990; Loeffler et al., 1993].

Another level of flexibility was found for lineage specification within the hematopoietic tissue. It is possible to bias the degree of erythroid, granuloid and macrophage lineage commitment by several maneuvers altering the growth environment. The present concept to explain the fluctuations observed in lineage specification is based on a dynamic network of interacting transcription factors involving the PU-1 and GATA molecules [Zhang et al., 1999]. Cross and Enver put forward the concept of fluctuating levels of transcription factors with threshold dependent commitment [Cross and Enver, 1997].

There is now a rapidly growing literature that tissue stem cells specified for one type of tissue (e.g. hematopoiesis) can be manipulated in such a way that they can act as tissue stem cells of another tissue (e.g. neuronal, myogenic) [Bjornson et al., 1999; Brazelton et al., 2000; Seale and Rudnicki, 2000; Goodell et al., 2001]. The major trigger that can redirect the tissue specification seems to be the growth environment into which the cells have to be placed. Very clearly this tissue plasticity represents a particular degree of flexibility consistent with the above definition. On the other hand this phenomenon makes it necessary to include into the above definition the growth environment by which the stem cell potential is restricted to a specific function. This supports the addition of the homing criterion into the amended definition. However, at present it is largely unclear whether tissue stem cells, reprogrammed for another tissue, are really fulfilling all criteria of our definition. It will be necessary in the future to carefully check all the requirements listed.

Whereas the term *stem cell plasticity* is often used to describe the potential of stem cells to alter tissue- or lineage characteristics, we introduce the term *within-tissue plasticity* to describe flexibility and reversibility of cellular characteristics restricted to one tissue or cell lineage.

In summary the amended definition of tissue stem cells that we propose gives the following features a greater emphasis than previous definitions:

- a shift from the cellular view to a system view including self-organizing processes;
- emphasizing stemness as a capability rather than as a cellular property;
- introducing the growth environment;
- emphasizing within-tissue plasticity;
- upgrading the functionality of the tissue stem cells and the tissue;
- extension of self-maintenance to self-renewing capability.

Implications of the Definition

The amended definition has several implications which we like to discuss.

Definition of Maturing and Transit Cells

Maturing cells can be defined as cells with full expression of functional differentiation markers, no capability of proliferation, no capability of self-renewal or self-maintenance, and hence, no ability to regenerate tissue after injury. Maturing cells therefore represent cell stages which are close to completing their development and becoming functional end cells. In this context, for example, reticulocytes would be maturing cells, as would be segmented neutrophils in the bone marrow, villus cells in the intestine and stratified cells in epidermis.

Transit cells can be defined as a cell stage which is intermediate between stem cells and maturing cells. We define transit cells by the following criteria: they are characterized by the onset of differentiation marker expression during their development which are, however, not mandatory, they are capable of proliferation, and they do not self-maintain or self-renew. This implies that transit cells may be able to operate as amplifying cell stages generating many maturing cells from the few cells entering the transit cell stage. Therefore, transit cells would be capable of producing many progeny cells (criterion (S4)) which are temporarily capable of regenerating a tissue after injury (criterion (S6)). However, no long-term regeneration and no functional re-establishment of the tissue would be possible.

Uncertainty Principle

A question frequently posed is: 'Is this cell a stem cell?' We refer to this as the *ontology question*. It implies the idea that one can decide about the capabilities of a given cell without relating it to other cells and without testing the capabilities functionally. We believe that this is a very dogmatic and unrealistic point of view.

As we have seen above, the main attributes of stem cells relate to their potential in the future. These can only be studied effectively by placing the population of cells in a situation where they have the opportunity to express their potential. Here, we find ourselves in a circular situation. In order to answer the question whether a cell is a stem cell we have to alter its circumstances, and in doing so inevitably lose the original cell, and in addition we may only see a limited spectrum of responses. This situation has a marked analogy with *Heisenberg's uncertainty principle* in quantum physics. In simple terms, this principle states that the very act of measuring the properties of a certain system inevitably alters the characteristics of that system, hence, giving rise to a degree of uncertainty in the evaluation of its properties. We believe that this analogy holds true for the functional tissue stem cells in a very fundamental sense. The system is made to react to perturbations. Therefore, it might be impossible to determine the status of a single stem cell without changing it. We hereby postulate the fundamental uncertainty property for tissue stem cells. This implies that one will not be able to make a definitive statement about whether or not a given cell is a tissue stem cell. It implies that all statements that we can make will be necessarily probabilistic statements about the future behavior under particular conditions. Essentially this has two particular aspects. The first is that we can only make statements about cell populations in the statistical sense of expected values under a given statistical model. This implies that measurements will necessarily be conducted on populations of cells. The second essential aspect is that we cannot disregard the experimental procedure by which the stem cell under consideration was challenged. Each assay procedure and each perturbation process may induce a different response in one or several of the characteristics of the tissue stem cell. This is constitutive to the stem cell property, as it is thought to be reactive and responsive to various types of perturbations. Hence, all statements about tissue stem cells and their reactions have to be given in the context of the perturbation of the measurement process under which they were obtained.

This type of uncertainty concept makes it obvious that the ontology question can be rather misleading.

Symmetric and Asymmetric Cell Division

A number of mechanistic models have been proposed in the past on stem cell growth and regeneration. Many of these, including some of our own work have introduced assumptions about symmetric and asymmetric stem cell divisions in the hematopoietic and intestinal tissue [Vogel et al., 1968; Loeffler and Wichmann, 1980; Loeffler and Grossmann, 1991; Loeffler et al., 1993; Loeffler et al., 1997]. All these concepts tightly link cell division with the decision on self-maintenance. We now consider that such concepts are too rigorous and potentially misleading. In one way or another they all require that the cells somehow explicitly 'know' how to behave. Furthermore, these cells need to inherit this knowledge to there daughter cells in either a symmetric or asymmetric way. Our definition does not imply any assumptions about the correlation of proliferation with self-maintenance or self-renewal. Hence, no implications about symmetric or asymmetric stem cell divisions enter.

Predictive Quantitative Stem Cell Models

The Need for Models

Although experimental research on tissue stem cells has attracted a lot of attention it is surprising how little theoretical work has been done. We only give a subjective list of important references on mathematical modeling: Till et al., 1964; Mackey and Glass, 1977; Loeffler and Wichmann, 1980; Ogawa and Mosmann, 1985; Grossman, 1986; Loeffler and Grossmann, 1991; Abkowitz et al., 1996. Most of these models (including many of our own) do not fulfill the complete criteria list mentioned above. We believe that there is a tremendous need for predictive quantitative theories and simulation model of tissue stem cells consistent with the above concepts. The lack of theory may be partly due to the expectation that one will be able to determine stem cells directly. The more we realize that this expectation is misleading we need concepts to cope with our lack of understanding. A theoretical basis of tissue stem cell functioning will have several advantages:

- Theories provide presumptive mechanisms to explain and link a variety of observed phenomena and reveal how far data are consistent with one another and with the latent mechanisms.
- Theories help to direct experimentation due to predictions that can be investigated.
- Theories help to anticipate the impact of manipulations to a system and its response. A situation that might be relevant in designing treatment strategies.

- Theories adapted for different tissues may help to understand the similarities of construction principles between tissues.

Subsequently we give a list of general requirements such models should fulfill in our mind.

General Model Requirements

Based on the preceding discussion we can deduce general requirement models of tissue stem cells must fulfill:

(M1) The model cells must fulfill the criteria (S1)–(S7). This has the following implications:

(M2) The models must be based on populations of individual cells to follow clonal development, to conform with the uncertainty principle and to enable considerations of population fluctuations.

(M3) The models must consider growth environments and the interactions with the cells.

(M4) The system has to be dynamic in time and possibly space.

(M5) The system requires assumptions on mechanism to regulate proliferation, cellular differentiation and cell – growth environment interactions (e.g. homing).

(M6) The model concept must be comprehensive in the sense of being applicable to the normal unperturbed in vivo homeostasis as well as to any in vivo or in vitro assay procedure. This criterion requests that system–measurement interactions must be consistently considered.

Some Ideas about Mechanisms

Any model on tissue stem cells has to specify assumptions about the mechanisms that control the properties proliferation, differentiation, maturation, lineage specification and homing. We have developed a novel view on how to choose such assumptions. This view radically differs from other concepts presented so far in the literature.

Our approach strictly avoids assumptions that end up with direct or indirect labeling of particular cells as stem cells a priori. We rather attribute to all model cells only functional properties and request that the system behavior selects some of the cells by virtue of the properties such that they fulfill the criteria of the definition.

This approach attempts to completely avoid explicit stemness properties. Hence, the task is to design a dynamic process that drives and controls the cellular attributes. The leitmotifs here are the aspect of capabilities (i.e. actual and potential expression of a property), of flexibility, and reversibility. Apparently these aspects are controlled by the genetic and epigenetic status of a cell and by the activity of the signal transduction pathways including the transcription factor networks. Clearly, it is presently impossible to describe these processes in any reasonable detail. It will therefore be necessary to propose a simplified basic scheme of cellular dynamics.

Let us consider activation of genes relevant for the behavior of TSC. There may be circumstances when activation is not possible. In order to make activation possible a number of conditions must be met. This includes e.g. epigenetic constellations [Bonifer, 1999] and the availability of molecular partners to form transcription factor complexes. We can therefore conceptually distinguish two levels of gene activity control. Level 1 is qualitative and decides whether a gene is accessible for activation or not (sensitive or insensitive). Level 2 is quantitative and describes the degree of gene expression in a sensitive gene. Active gene expression may be high or low but can also be zero. In this two level control concept a gene may not be expressed for two very different reasons. It may either not be sensitive (level 1 dynamics) or it may be sensitive but there is no activation due to lack of challenge (level 2 dynamics).

State-transition graphs can now be classified. If they contain only self-maintaining and irreversible acyclic transitions between states, a population can be self-maintaining but not self-renewing. Irreversible transitions are associated with a gradual drift of the population to some selected states. In figure 1B long-term development will eventually reach a composition of the population with very few cells having active gene B. In panel A the state transition graph is characterized by reversible transitions. This would imply the property of self-renewal.

The second level of dynamics takes place in each of these states. Let us assume two genes (properties A and B) that can be activated. If we follow the pathway of activation over time we can plot this in a state-space diagram (fig. 2).

Furthermore, we assume that the dynamic development of gene expression depends on the growth environment a cell is homing to. Figure 3 illustrates how this can affect the preferred direction of development. If a cell could be shifted from one growth environment to another the drift can be changed. Hence, alternating homing to various growth environments would yield a rather fluctuating trajectory. In such a setting not only the influence of the environments would be considerable but in particular the frequency of transitions between them.

Taken together such a general concept of two-layered property dynamics would be a possibility to build into tis-

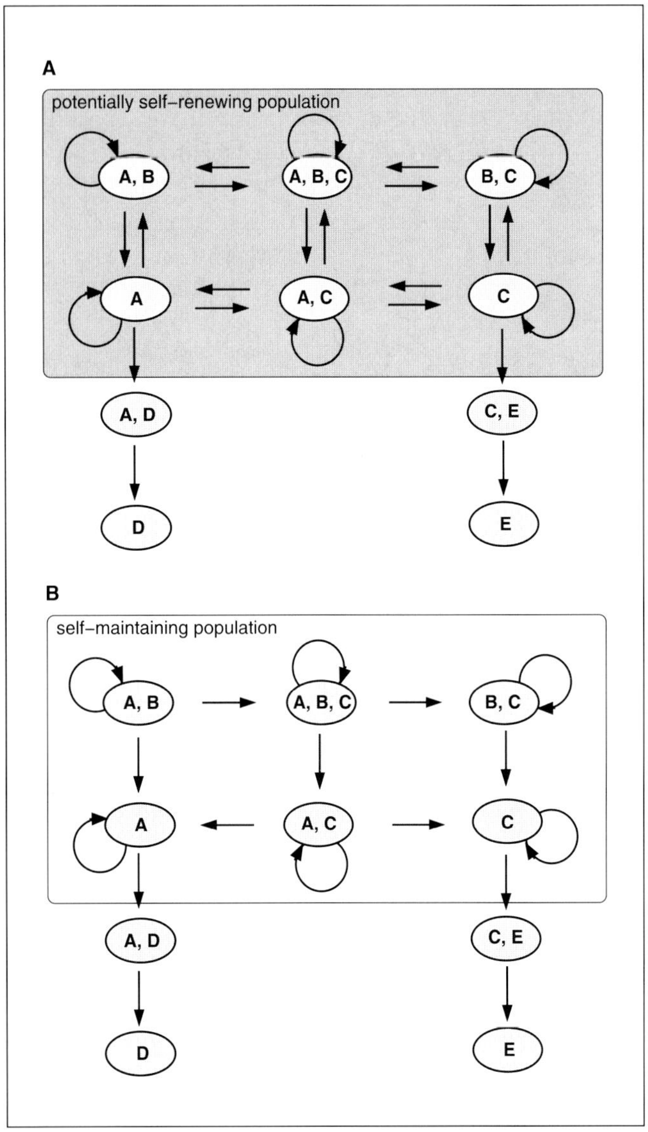

Fig. 1. State transition graph for self-renewing and self-maintaining population. Provided is a diagrammatic sketch of a level 1 dynamic to illustrate the difference between the criteria of self-renewal (**A**) and self-maintenance (**B**). We assume that the genes A-E can be made accessible for activation (sensitive). Tuples of permitted combinations of sensitive genes define possible states. Possible transitions between some of these states are indicated by arrows. This leads to the description of a state-transition graph for the qualitative dynamics present on level 1. A transition involves loss or gain of a sensitive gene.

sue stem cell models. We subsequently indicate how this can be implemented giving two examples.

Modeling the Hematopoietic Stem Cell System

In the following paragraph we will describe the basic ideas of a model concept for the organization of the hematopoietic stem cell system which we propose [Roeder and Loeffler, 2002].

In this new concept we suggest a separation of potential cellular properties and their actual use. Expression and usage of these properties develop according to a specific cellular propensity which itself depends on the growth environment. To formulate the model we make the following minimal set of assumptions: (1) We assume two different growth environments inside the bone marrow (GE-A and GE-Ω) where stem cells are either attached to (GE-A) or detached from (GE-Ω) specific stroma components. (2) Each cell is characterized by two properties, the cycling status *c* (position in cell cycle) and a property *a*, which describes the affinity of the cell to reside in GE-A. (3) Residing in GE-A a cell is assumed to be non-proliferating (in G_0). In contrast, cells in GE-Ω proliferate with an average turnover time. (4) Whereas cells in GE-A have the propensity to increase affinity *a*, cells in GE-Ω tend to decrease *a*. (5) Cells can change from GE-A to GE-Ω and vice versa with intensities (probabilities per time interval) α and ω, which depend on the actual affinity *a* and the cell numbers in the system. (6) Cellular development with respect to properties *a* and *c* is reversible (within-tissue plasticity). Only if the attachment affinity *a* has become critically small the cell is considered to have lost the potential to stick to GE-A. Such cells we call differentiated. Figure 4 gives a schematic representation of the model.

This stochastic, single cell-based model meets the requirements listed as (M1)–(M6), including all criteria of the amended stem cell definition, and is predictive for cell populations as well as for single cell and individual clone dynamics.

The model stem cells are individual members of a heterogeneous (with respect to attachment affinity *a* and cycling status *c*) population of undifferentiated ($a > 0$) cells (definition criteria (S1)). They are able to home to an appropriate growth environment (S2). Here, we assume two different growth environments, each one inducing different developmental options. The cycling activity in GE-Ω ensures proliferation (S3). Repeated cycles of (i) cell division in GE-Ω, (ii) transition to GE-Ω, (iii) regaining *a* and (iv) back transition to the proliferative GE-Ω realizes cell production and therefore potential expansion

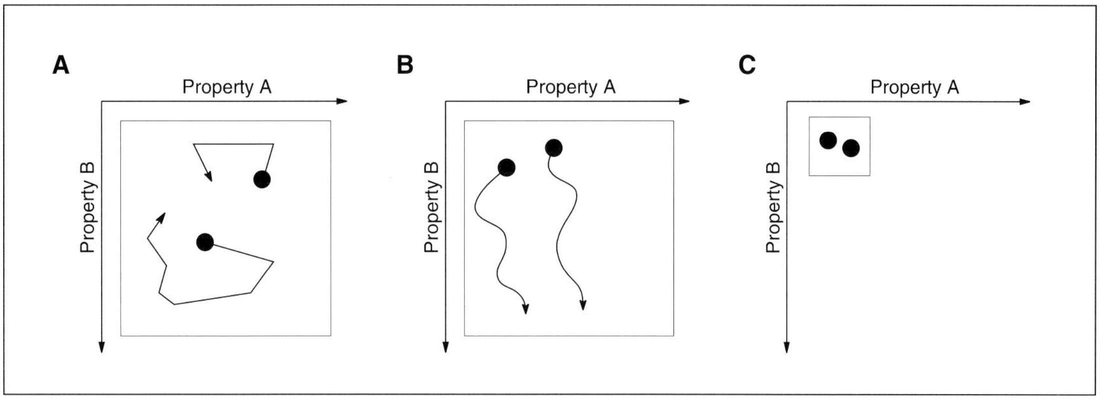

Fig. 2. Cellular development with respect to properties A and B. Shown are possible trajectories of two cells (●). In panel **A** they would fluctuate in a reversible way and would span a wide range of options (large within-tissue plasticity). In panel **B** this process would not be reversible in property B and hence, the scope of potential developments would be restricted (reduced within-tissue plasticity). **C** represents a minimal developmental potential (no within-tissue plasticity).

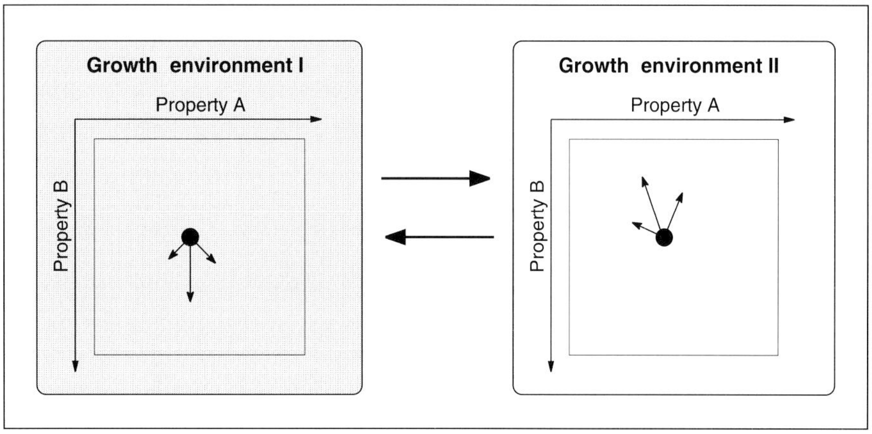

Fig. 3. Dependency of cellular development from growth environment. This figure illustrated the position of a cell (●) and the preferred developmental directions (arrows) with respect to properties A and B depending on the actual growth environment. Alternation between different growth environments could induce fluctuating expression of cellular properties.

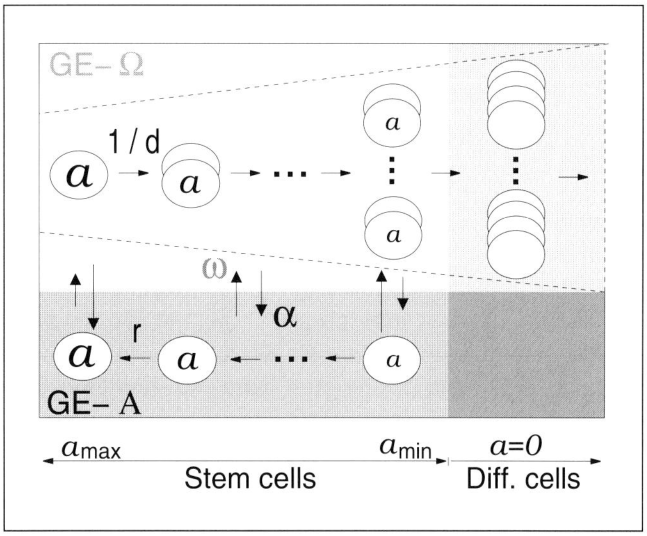

Fig. 4. Schematic representation of the model concept for hematopoietic stem cell organization. The lower part represents growth environment (GE)-A and the upper part GE-Ω. The dashed trapezium illustrates cell amplification due to proliferation. Depending on the assumed turnover times, different cell numbers will be produced (cell groups separated by vertical dots). Whereas attachment affinity a decreases by factor $1/d$ per time step in GE-Ω, it increases by factor r per time step in GE-A. The actual quantity of a is sketched by different font sizes. If a has fallen below a_{min} the cell has lost its potential to switch to GE-A (represented by empty cells: $a = 0$) and is called differentiated. Transition between GE-A and GE-Ω occurs with intensities α and ω.

of the stem cell population in its original composition. This process guarantees the self-renewal of the stem cell population (S5) and, together with the release of cells from GE-Ω (if $a < a_{min}$), the production of a large number of differentiated, functional cells (S4). Furthermore, the model is able to fully re-establish the system from one cell and to compensate repeated damages dynamically (S6). The fact that all members of the stem cell population are able to perform different functionalities (e.g. proliferation, regain of *a*, initiation of differentiating clones) clearly points out the requested flexibility (S7) of such cells.

Because the model is using single cell-based simulation procedures it is able to describe single cell as well as cell population behavior. This includes the possibility of tracking individually marked clones. The realization of the self-renewal and the cell production process is inevitably linked to the existence of two different growth environments. These, however, should not be interpreted in a spatial sense. More important is the existence of a different signaling context. The process of transition between the two growth environments is dependent on individual properties of the cells (attachment affinity *a*) and on the cell numbers in the system. Therefore, both, the composition (with respect to *a*) and the size of the cell population act as feedback mechanisms for the dynamic regulation of the system. In our approach we assume that the attachment process of stem cells to stromal elements is a central requirement for the induction of signaling processes which than enable the cells to activate different developmental programs. The model is able to produce qualitatively different growth scenarios. As mentioned earlier, this includes the capacity to fully re-establish the system from one cell and to compensate repeated damages. Steady state situations are dynamically stable and robust against perturbations. Depending on the model parameters the system behavior differs qualitatively. Besides the steady state, situations of system exhaustion and cycling behavior are possible.

Using simulation studies we can show that the model is able to describe a huge variety of experimental settings and different kinds of in vivo and in vitro assays. This includes *fluctuations* in the contribution of differently marked but otherwise identical cell populations [Abkowitz et al., 1996] as well as the *dis- and reappearance* of kinetically different cell types in mouse chimeras [Van Zant et al., 1992]. But not only the description of fluctuating behavior on the level of cell populations is possible. Our model also reproduces experimental results of *individual clone tracking* experiments. Strikingly, contradictory observations [Jordan and Lemischka, 1990; Drize et al., 1996] can be explained consistently assuming different sampling techniques and detection thresholds. The *dependency of stem cell development on the growth environment* is also included in the model. So it is able to explain differences in engraftment potential of stem cells cultured on different kinds of stroma cell lines including the effect of direct stem cell–stroma contact [Wineman et al., 1996]. The *variability with respect to clone size, clonogenic potential or time to clonal appearance,* which has been reported for various colony growth assays [e.g.Till et al., 1964; Hao et al., 1996] can be explained by our concept without assuming a predefined heterogeneity of the stem cell population. As a last example of the variety of experimental results covered by the model, we mention the *characteristics of continuous S-phase* labeling with bromodeoxyuridine (BrdU). The model is consistent with the observation that, although the majority of stem cells is non-proliferating under steady state conditions, most of the primitive cells can be labeled in S-phase within a couple of weeks [Bradford et al., 1997; Cheshier et al., 1999].

Similar to other stem cell models our approach is also based on a stochastic decision of individual cells. But in contrast to the assumption of different kinds of cell divisions (symmetric/asymmetric) which are chosen with predefined probabilities [Till et al., 1964; Vogel et al., 1969], in our model the stochastic nature of cell development relies on dynamically controlled switches between growth environments. The stochastic nature of this process seems to be appropriate because it is assumed to be influenced by a highly complex signaling network, which cannot be specified in detail. Other models, such as one previously proposed by our group [Loeffler and Wichmann, 1980], allows a dynamic regulation of the self-renewing probability by tissue-specific feedback mechanisms. However, this model purely described populations of cells and was not able to investigate single cell fates and therefore clonal development. The same restriction appears in the clonal succession model of Abkowitz et al., 1996 which suggests a stochastic model where a self-renewing stem cells pool successively releases clone-initiating cells. Stem cell–environment interactions are also not included in this model. This is different in the concepts proposed e.g. by Schofield [1978], who introduced specific stem cell niches or by Muller-Sieburg and Deryugina [1995], who linked the binding of stem cells with stroma components to a differentiation arrest. Both concepts suggest mechanisms which regulate proliferation and self-maintenance, however, they explicitly exclude reversibility in the development of the stem cells. We think that this assumption has to be questioned in the

Table 1. Phenomena explained by hematopoietic stem cell model

Phenomena	Assay/data
Clonal fluctuation	
Cell population level	Competition assays[1]
Individual clone level	(Retro-)viral marking experiments
Competition of different cell types	Competition assays[2]
Conversion to monoclonality	*model prediction, no experiment known*
Microenvironmental dependence	Co-culture on stroma cell lines/transwells
Heterogeneity of stem cell population	
Clonogenic potential	CFU-S assay
Clone size	CFU-C assay
Time to clonal appearance	CAFC assay
Cycling activity	Continuous BrdU labeling

1 Using differently marked but otherwise equal cell populations.
2 Using (cell kinetically) different cell populations, e.g. DBA2/C57BL6, mouse strains.

light of the recent experimental results on stem cell plasticity. The development of our model concept has been influenced by these and other ideas which cannot be listed here completely, however, it is extended to a microscopic mechanism explaining the dynamics of stem cell self-renewal, allowing for the tracking of individual clones and offering a explanation for reversibility and plasticity phenomena.

Modeling the Intestinal Epithelium

We and others have previously undertaken extensive modeling efforts to characterize the cellular program of cell proliferation and differentiation in the intestinal crypt system [Potten and Loeffler, 1987a; Paulus et al., 1993; Meineke et al., 2001]. These models used the concept of specialized stem cells undergoing mostly asymmetric self-maintaining cell divisions and a pedigree development of transit cells which undergo a predetermined number of cell cycles before they enter the status of non-proliferative maturation. They are very successful in explaining spatial and cell kinetic data during steady state homeostasis. Data on positional mitotic and label indices, on the migration of cells to higher crypt positions, on the shape of clones of mutated cells, on cell turnover rates, on the generation of goblet cells in mid crypt positions and data on localizations of apoptotic cells can be explained by cellular automata models. They describe the behavior of single cells arranged on a 2-dimensional flask-shaped topology corresponding to a single-layered epithelium in a jejunal or colonic crypt.

Problems, however, arise with the concept of specialized asymmetrically dividing stems cells, transit cells, and pedigree development in two situations deviating from homeostasis. One situation concerns the long-term behaviour of intestinal crypts. There is increasing evidence that crypts duplicate by crypt fission, a process in which a spatial separation starts at the bottom of a crypt where the presumptive stem cells are located [Totafurno et al., 1987]. This implies that the stem cells of a crypt have to be shared by two daughter crypts. Obviously this implies a process of amplifying the number of tissue stem cells in crypts to enable colonization of the daughter crypts. Furthermore, there is evidence that the population of stem cells actually active in a crypt is undergoing fluctuations. These can be observed indirectly by mutagenesis experiments which permit to follow the occurrence of clones of cells carrying a mutation in a staining property. These clones can be followed over time and data show that in the long run all crypts convert to monoclonality [Winton and Ponder, 1990; Park et al., 1995]. Hence, either all cells in the crypt carry the mutation or the mutation is completely lost from the crypt. We have undertaken extensive modeling on this problem [Loeffler and Grossmann, 1991; Loeffler et al., 1993; Loeffler et al., 1997] and conclude that a certain fraction of actual stem cells in a crypt is lost every day and substituted by others which are recruited nearby. This process appears to be controlled. Hence, these results indicate that at least the population of actual stem cells in a crypt is not static but subject to continuous fluctuations. It is, however, not clear whether there is a pool of dormant potential stem cells which can be recruited or whether transit cells can be recovered to act as actual stem cells. Indications for the latter come from a second piece of evidence. There are extensive investigations on the regeneration of the intestinal crypt after a severe radiation injury [Wright and Alison, 1984; Potten, 1991]. Typically, after radiation cell production ceases for a few days and the cell number drops rapidly. After about 3 days regeneration starts and results in a very rapid expansion of crypts with overshooting cell numbers within one week. When we analysed these dynamics using a model approach we had to conclude that this rapid and overshooting behaviour could only be explained if one assumed that transit cells would suddenly behave like tissue stem cells in order to increase cell production [Paulus et al., 1992]. The number of actual stem cells believed to exist from steady state growth is by far insufficient to generate the huge cell production going on after this injury. Furthermore, data on clonogenicity experiments using split dose radiation also provided strong evidence that a crypt contains many more potential stem cells than are actually participating in a homeostatic situation [Roberts et al., 1995; Berger et al., 2001].

These findings lead us to seek for alternative stem cell models for the intestinal crypt. The basic idea is to give up the concept of explicitly programmed stem cells completely and to consider only a set of properties attributed to cells whose actual status is controlled by a growth environment. The growth environment we introduced in the new model is a gradient of a stimulatory substance diffusing from the bottom of the crypt to the top. We were in fact inspired by observations of Bjerknes et al. [1985] about the existence of intraepithelial gap junctions that would permit the passage of molecules from columnar cell to columnar cell. We assume that the growth factor degrade with time. This leads to a gradient of growth factor concentrations from the bottom of the crypt to the top. A first model investigating the potential of this concept in a one-dimensional chain of concatenated cells was very convincing [Gerike et al., 1998]. It proved that one can fully replace the old assumptions of a pedigree and dedicated stem cells if one introduces only one rule on the control of cell proliferation. We assume that cells in which the growth factor concentrations exceed a certain threshold continue to divide while a concentration below the threshold leads to a proliferation stop. As the previous models this concept provides a very simple and consistent explanation on all cell kinetic data, but in addition it provides an elegant explanation of the recovery after damages as e.g. radiation injury [Gerike et al., 1998]. The major novelty in this model is that all proliferative cells are alike and all can potentially contribute to tissue regeneration. In particular there is no explicitly nominated stem cell in this system.

Two-dimensional modeling of this concept is under way (fig. 5). All results we have so far indicate that the novel concept permits a consistent explanation of several other phenomena. Even though all proliferative cells are considered to have the same potential to generate clones only few cells will actually be able to establish long living clones. These are cells which happen to be trapped in a spatial niche formed by the rough landscape of non-proliferative paneth cells at the bottom of the murine jejunal crypt. Although not rigidly fixed in these niches the cells find themselves in a place of little flow and hence they are not easily washed away. Therefore, they can give rise to clonal expansion of their progeny for a long time period. However, in the long run it happens that either the cells are eventually washed away by some neighbor cells or the niche itself changes spatially. As a consequence the clone-producing cell disappears and others are successful in finding such a niche. Clearly, if one eliminates proliferative clone-producing cells from this tissue (e.g. by damage) other proliferative cells located near the crypt bottom can immediately take over. Thus we can show that this kind of model provides a straightforward and natural explanation of a wide range of phenomena as coexistence of several actual clone-producing cells at any point in time, long-term fluctuations in the clonal composition in each crypt, long-term conversion to monoclonality with regard to any marker introduced in the past or resistance of the crypts to damage (radioresistance) due to the large number of potential stem cells capable of starting regeneration. It should be mentioned at this stage that similar model results would be obtained for any mechanism that produce a spatial cut off of factors stimulating proliferation. It will be interesting to see whether the neuron-mediated GLP-2 growth factor-based control process recently unrevealed by Bjerknes and Cheng [2001] satisfies such properties.

Table 2 summarizes the phenomena explained by the novel class of model so far. Publications on these results are in preparation.

Model Comparison

Let us consider how these models of hematopoietic and epithelial stem cell systems relate to one another and how they correspond to the novel paradigm of tissue stem cells.

Both models prove that one can conceive regenerative tissue systems fully consistent with the criteria of the

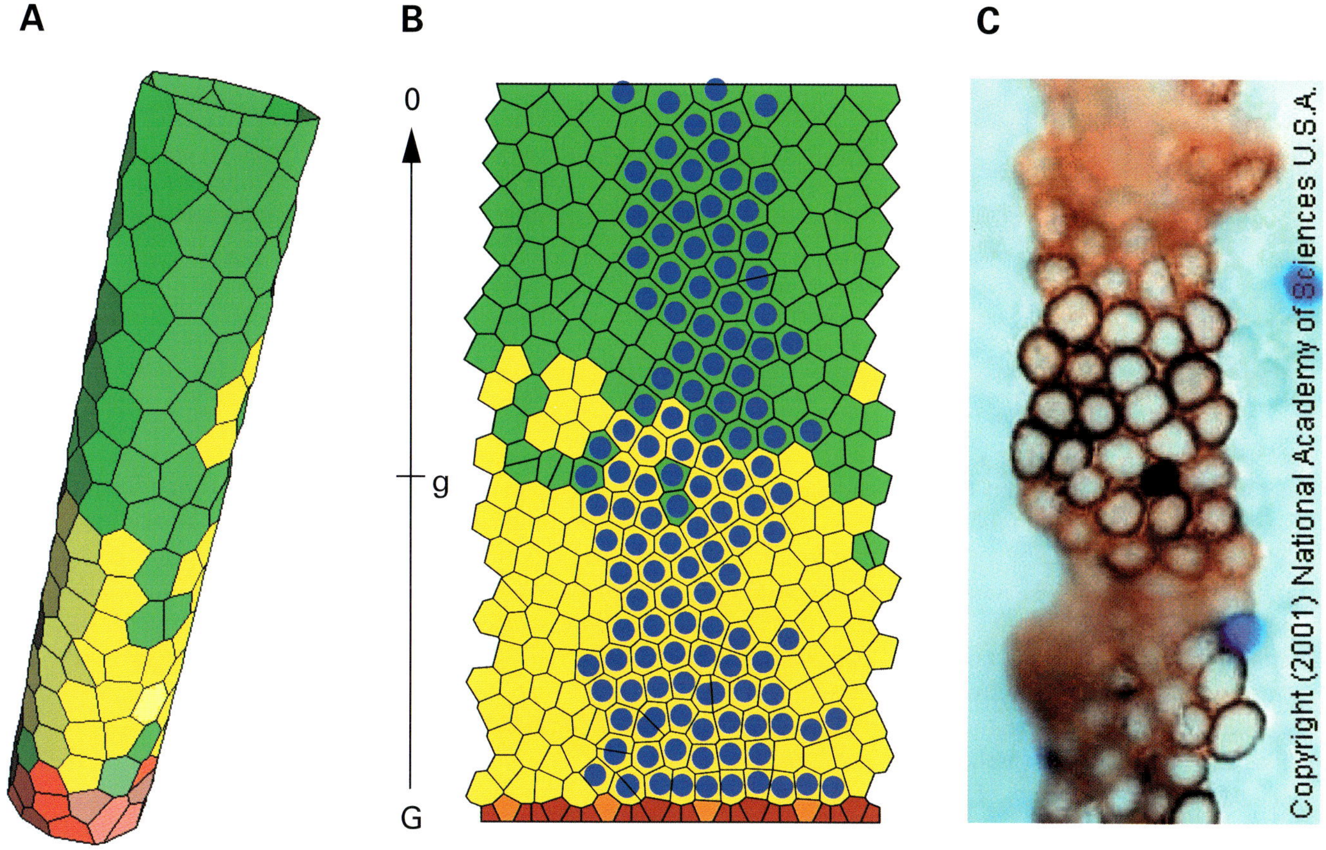

Fig. 5. Two-dimensional crypt model. **A** Cylinder model, representing the crypt structure. **B** Unrolled cylinder. Paneth cells are shown in brown, proliferating cells in yellow and non-proliferating cells in green. A growth factor G is produced at the bottom of the crypt and diffuses through cells establishing a factor gradient. Only cells with growth factor concentration above threshold g proliferate. Cells which are derived from one individually marked proliferating cell are indicated by blue dots. At a later timepoint this clone might disappear or populate the entire crypt. **C** Experimentally observed cell clone on crypt/villus tissue 3 weeks after applying a stainable genetical mutation to one cell [from Bjerknes and Cheng, 2001].

Table 2. Phenomena explained by crypt stem cell model

Phenomena	Assays/data
Steady state regeneration	
Spatial arrangement	Data on positional distribution of cell types
Cell kinetics	Positional mitotic and label indices
Cell production	Migration velocity of columnar cells
Pluripotency	Production of goblet cells
Coexisting of clones	Mutation experiments (short term)
Conversion to monoclonality	Mutation experiments (long term)
Regeneration after damage	
Decline of cell numbers	Post radiation cell kinetics
Rapid start of regeneration	Post radiation cell kinetics
Radioresistance	Tissue dose response curve

Table 3. Comparing tissue stem cell models

	Hematopoiesis	Intestinal crypt
Controlled cellular properties (potentials)	$G_0 \leftrightarrow$ in cycle Affinity to GE-A	Proliferation $\rightarrow$ maturation
Controlling mechanisms	Two growth environments with different impact on cellular properties (within-tissue plasticity)	One growth environment for the maintenance of cellular properties
Major source of functional heterogeneity	Stochastic switches between growth environments	Stochastic fluctuations in spatial cell arrangements

amended stem cell definition, leaving aside assumptions on pedigrees, preprogrammed asymmetric divisions or other assumptions implying a priori the entity of predetermined tissue stem cells. Both models are based on the idea of cells which have properties subject to regulation. It is then the interaction of the cells with one another (e.g. competition for growth environments or spatial niches) and with the growth environments which determine the actual development and whether a particular cell may act as an actual clone producing cell. Nevertheless the models differ in the specification of these assumptions. Table 3 gives a summary.

In the hematopoietic model two cellular properties are subject to control: (1) Whether a cell cycles actively or not and (2) whether it increases or decreases the affinity to growth environment GE-A. In the intestinal model we only assume one controlled property i.e. the decision of whether a cell continues proliferation or is converted to the non-proliferative maturing state. This decision is deemed to be eventually irreversible. There are also differences regarding the growth environments. Our hematopoietic model is based on the idea of two growth environments between whom cells can alternate. Thus fluctuations enter primarily due to the stochastic nature of this switching process. This basically generates a within-tissue plasticity in the model. The model provides an explanation for the continuous generation of a cellular heterogeneity in the population of stem cells and links this with a process of self-renewal. The intestinal model so far requires only one growth environment hypothetically related to a growth factor gradient. The model assumes a homogenous population of proliferative cells all of which are capable to act as actual stem cells if they are placed into a spatial niche at the bottom of the crypt. Hence, in this model stem cell heterogeneity is not considered but the spatial restrictions only permit a small subset of potential stem cells to contribute to tissue formation at any point in time. Fluctuations in the clonal composition would originate in this model not from a switch between growth environments but due to fluctuating spatial positioning of cells within this growth environment.

Discussion

A major point in this conceptual paper was to provide a critical discussion of the definition of tissue stem cells. Only few papers deal with the conceptual aspects of this matter despite an expanding set of biological data. We have revised a definition put forward by Potten and Loeffler [1990] in the light of new conceptual considerations and experimental findings. The major conclusion is that a functional definition of tissue stem cells still is the baseline.

Tissue stem cells are conceived as cells capable of behaving in a variety of ways and hence, it is their potential and the flexibility to use this potential that matters. We argue that it is conceptually misleading to consider stemness as a specific property that can be determined at one point in time without putting the cells to functional tests. The potential of stem cells rather relates to the complexity of the state transition graphs describing the potential dynamics of gene activation than to the actual activity status in one of these states.

This has implications for attempts to define tissue stem cells e.g. by gene profiling [Phillips et al., 2000]. There are several problems that we envisage. First, gene profiles obtained by microarray technology are measured on cells obtained from negative selection procedures leading to a heterogeneous mixture of cells. Second, the assays typically represent snapshots at one point in time. However, such snapshots give little insight into the potentials of a stem cell population. It would be essential to track the gene profiles over time in various experimental settings putting the tissue stem cell system under various modes of stress. This will be necessary to sketch the topology of the gene activity network and its reversibility. Thirdly, to conform with the functional definition it will be crucial to correlate the gene activity network to the functional capabilities of the cells in functional assays. Hence, all techniques based on snapshot measurements of some surface markers or gene activity patterns must be considered as surrogate techniques.

At present we cannot see the possibility for a molecular definition of tissue stem cells disregarding functional aspects as a reference point. Thus, we are reluctant to believe that tissue stem cells can be defined by a 'tissue stem cell chip'. Such an approach would basically ignore the two basic aspects of stem cell potentiality and of cell–growth environment interaction. Furthermore, the uncertainty problem would still apply and all statements could only be made in a probabilistic sense. However, we expect from gene profiling possibilities to select cells with properties required for stem cells and one can expect a more detailed insight into the mode of stem cell operation by investigating the underlying mechanisms. In particular one can hope for test procedures to screen functional capabilities of tissue stem cells.

The concept proposed above changes the paradigm in thinking about stem cells. Rather than to think about these cells as being specialized in the first place, we suggest that they are selected and modified due to interactions with the growth environment. Their properties are considered to permanently fluctuate so that some cells meet a situation of expansion and growth.

We are in fact able to demonstrate by computer simulations for at least for two tissues (hematopoiesis and intestinal epithelia) that this kind of tissue self-organization is consistent with the above stem cell definition. This modelling also shows that we must be careful with our statements about the way stem cells actually operate. The model mechanisms proposed by us have at least the same strength as pedigree models proposed previously. It must be admitted that our own group for many years has been a strong proponent of the classical stem cell and transit cell pedigree concept [Potten et al., 1982; Loeffler et al., 1987; Potten and Loeffler, 1987b; Wichmann et al., 1988; Loeffler et al., 1989; Paulus et al., 1992]. Several investigations are now underway to compare the two classes of theories but at present the novel theory appears superior in covering a much broader range of phenomena.

Generally we believe that much more work needs to be done to develop a comprehensive quantitative and predictive theory of tissue stem cells. Such theories in combination with simulation models are required to better understand the process of dynamic self-organization. The criteria for such models are listed above. A major challenge is to incorporate into these models mechanisms of lineage specification. Both, the hematopoietic and the intestinal cell systems, produce several cell lineages. If we follow the reasoning of Cross and Enver [1997], the basic idea to investigate is a fluctuation on the level of transcription factor concentrations which generate a kind of noisy ground state containing all options. Commitment into one lineage would then be a process of restricting the options (i.e. loss of some properties) and limiting reversibility. The gene activity network shown in figure 1 gives a sketch of how two lineages might originate. Thus we anticipate a straightforward generalization of our model class to include lineage specification within a given tissue environment.

There are a number of predictions arising from these models already. One basic prediction is that two twin cells originating from the same mother cell put into different growth environments will take different developments paths. This is, however, also predicted if they are placed into identical growth environments. The ongoing fluctuations will eventually lead to different fates.

Another prediction concerns clonal evolution. Our model simulations of hematopoiesis and the intestinal crypt are based on a simultaneous activity of several coexisting tissue stem cells. They generate several clones and the situation is polyclonal at any given point in time. This should be evident always shortly after introducing some genetic markers (e.g. retro- or lentiviral marking). However, there are fluctuations and some active stem cells become silent (or get lost) and others are activated. Thus the clones contributing to tissue formation change with time. Actually in the long run the pattern is predicted to change. If one could label all cells in a tissue with a unique marker our simulations would predict that coexistence is impossible in the long run and that decedents from one clone will eventually generate all active stem cells in the tissue (crypt, hematopoiesis). This conversion to long-term monoclonality is a consequence of fluctuations. It would, however, not be possible to know in advance which clone will be the winner. Hence, we predict that, depending on the time scale of measurement, it is equally valid to argue that stem cell systems are polyclonal (actual activity) and monoclonal (descendent status) at the same time. A detailed understanding of the long-term dynamic features will be important in gene therapy based on random insertion of genes into tissue stem cells.

A third important model prediction concerns the role of self-renewal. If one has a stem cell system with a homogenous population of cells, self-renewal and self-maintenance are actually equivalent. In stem cell systems with heterogeneity the distinction is very important (see fig. 1). One can prove that systems which are only capable of self-maintenance can live for a long time but will with certainty die out at some point in the future. The reason is that once a sub-population at the root of the network is lost (e.g. in fig. 1B) it cannot be recovered. Other network nodes may substitute the situation but they also can get lost. Self-renewal is therefore a mandatory prerequisite for a system that is structurally robust against repeated damage and extensive stress. We therefore predict that self-renewal is an essential property of a stem cell system, but it may be a very slow and selective process and therefore difficult to detect.

The question if organismal age has an effect on the reversibility and flexibility potential, and therefore also to self-renewal of stem cells cannot be answered at the moment. It is unclear if observed differences in repopulating efficiency of stem cells from different sources (e.g. fetuses, young or aged adults [Rebel et al., 1996; Chen et al., 2000]) are caused by (irreversible) changes of the cellular potential or if the potential is simply not used due to a shift in the signaling context.

These aspects raise the question of appropriate assays for self-maintenance and self-renewal. We here refer to a previous discussion in Loeffler and Potten [1997]. Self-maintenance can be determined by serial transplantation experiments in vivo and in vitro. Basically one investigates whether clonogenic cells can be recovered and expanded repeatedly. Surrogate strategies avoiding transplantations use repeated damage-recovery cycles in intact tissues or quantifications of cell numbers produced by cell clones. These techniques are all not very accurate and depend on the experimental procedure and on the timing of the read out. Assaying self-renewal is even more difficult as it also requires to investigate the composition of the cell populations to make sure that not only the overall population but also its composition with respect to subpopulations recover. At present there are no good assays to do this. Gene profiling may become helpful, if one can show that a biased pattern after damage can recover to a control pattern given time and appropriate conditions. These will be important experiments to undertake.

Our reasoning has emphasized the role of cell–growth environment interactions. This implies that more attention needs to be paid to the role of the microenvironment. This is actually a complex subject itself. Growth environments encompass an element of spatial neighborhood to other stem cells and matrix cells, ways to adhere to them, and ways to receive signals (growth factors, direct cell contacts, gap junctions, pseudopods). Growth environments may home a cell for a certain while and can then be called a niche. However, such niches may have limited life times but little is known about the dynamic changes of growth environments. Any kinetic changes present will however increase the fluctuations in the stem cell population.

In this paper we did not have the objective to discriminate tissue stem cells from embryonic stem cells (ESC). It will be important to provide a definition of embryonic stem cells and contrast it with the one we have suggested for tissue stem cells. From our perception ESC are not self-maintaining or self-renewing as the embryo continues to develop. We speculate that the natural growth environment prevents this option. However, when taken into cell culture ESC can self-maintain very well [Robertson, 1987; Keller and Snodgrass, 1999] and placed into tissue-environments can generate tissue stem cells [Evans and Kaufman, 1981]. If the latter happens they have to fulfill the criteria given for TSC. Whether TSC can reversibly be manipulated to generate cells capable of acting as ESC in embryos is an interesting question which has to be solved in the future.

In summary, we proposed here a revised definition of tissue stem cells and discussed the conceptual and experimental consequences. We further discussed the need for quantitative theories of the self-organizing stem cell systems and presented outlines for two predictive simulation models applied to hematopoiesis and intestinal epithelia.

Acknowledgement

This work was supported by grants of the Deutsche Forschungsgemeinschaft DFG Lo 342/4-3 and Lo 342/9-1.

We are particularly thankful to Frank Meineke, Hilmar Berger, Dirk Drasdo, Michael Cross (Leipzig, Germany), Gerald de Haan (Groningen, NL), Catherine Booth and Christopher Potten (both Manchester,UK) for many inspiring discussions.

Furthermore we acknowledge the contributions of the participants (given in alphabetical order) of the 'Workshop on Tissue Stem Cells' held in Leipzig, Germany on June 12–14th, 2001: Zvia Agur, Malcolm Alison, Matthew Bjerknes, Constanze Bonifer, Dominique Bonnet, Catherine Booth, Michael Cross, Gerald de Haan, Christoph Engel, Tariq Enver, Myrtle Gordon, Ursula Just, Toshihiko Komatsuzaki, Brian Lord, Albrecht Müller, Christa Müller-Sieburg, Emanuel Nečas, Robert Oostendorp, Peter Quesenberry, Susanne Reif, Yutaka Saikawa, Olga Smirnova, Gary VanZant, Christof von Kalle.

References

Abkowitz, J.L., S.N. Catlin, P. Guttorp (1996) Evidence that hematopoiesis may be a stochastic process in vivo. Nat Med *2:* 190–197.

Abkowitz, J.L., D. Golinelli, D.E. Harrison, P. Guttorp (2000) In vivo kinetics of murine hemopoietic stem cells. Blood *96:* 3399–3405.

Anderson, D.J., F.H. Gage, I.L. Weissman (2001) Can stem cells cross lineage boundaries? Nat Med *7:* 393–395.

Berger, H., D. Drasdo, F.A. Meineke, M. Loeffler (2001) Post-irradiational small intestinal crypt survival can be explained by a pedigree free model. Cell Prolif *34:* 159 (abstract).

Bjerknes, M., H. Cheng (1999) Clonal analysis of mouse intestinal epithelial progenitors. Gastroenterology *116:* 7–14.

Bjerknes, M., H. Cheng (2001) Modulation of specific intestinal epithelial progenitors by enteric neurons. Proc Natl Acad Sci USA *98:* 12497–12502.

Bjerknes, M., H. Cheng, S. Erlandsen (1985) Functional gap junctions in mouse small intestinal crypts. Anat Rec *212:* 364–367.

Bjornson, C.R., R.L. Rietze, B.A. Reynolds, M.C. Magli, A.L. Vescovi (1999) Turning brain into blood: a hematopoietic fate adopted by adult neural stem cells in vivo see comments. Science *283:* 534–537.

Blau, H.M., T.R. Brazelton, J.M. Weimann (2001) The evolving concept of a stem cell: entity or function? Cell *105:* 829–841.

Bodine, D. M., P.S. Crosier, S.C. Clark (1991) Effects of hematopoietic growth factors on the survival of primitive stem cells in liquid suspension culture. Blood *78:* 914–920.

Bonifer, C. (1999) Long-distance chromatin mechanisms controlling tissue-specific gene locus activation. Gene *238:* 277–289.

Bradford, G.B., B. Williams, R. Rossi, I. Bertoncello (1997) Quiescence, cycling, and turnover in the primitive hematopoietic stem cell compartment. Exp Hematol *25:* 445–453.

Brazelton, T.R., F.M. Rossi, G.I. Keshet, H.M. Blau (2000) From marrow to brain: expression of neuronal phenotypes in adult mice. Science *290:* 1775–1779.

Cairnie, A.B., L.F. Lamerton, G.G. Steel (1965) Cell proliferation studies in the intestinal epithelium of the rat. I. Determination of the kinetic parameters. Exp Cell Res *39:* 528–538.

Cancelas, J.A., W.L. Koevoet, A.E. de Koning, A.E. Mayen, E.J. Rombouts, R.E. Ploemacher (2000) Connexin-43 gap junctions are involved in multiconnexin-expressing stromal support of hemopoietic progenitors and stem cells. Blood *96:* 498–505.

Chen J., C.M. Astle, D.E. Harrison (2000) Genetic regulation of primitive hematopoietic stem cell senescence. Exp Hematol *28:* 442–450.

Cheshier, S.H., S.J. Morrison, X. Liao, I.L. Weissman (1999) In vivo proliferation and cell cycle kinetics of long-term self-renewing hematopoietic stem cells. Proc Natl Acad Sci U S A *96:* 3120–3125.

Cross, M.A., T. Enver (1997) The lineage commitment of haemopoietic progenitor cells. Curr Opin Genet Dev *7:* 609–613.

de Haan, G., B. Dontje, C. Engel, M. Loeffler, W. Nijhof (1995) The kinetics of murine hematopoietic stem cells in vivo in response to prolonged increased mature blood cell production induced by granulocyte colony-stimulating factor. Blood *86:* 2986–2992.

Drize, N., J. Chertkov, E. Sadovnikova, S. Tiessen, A. Zander (1997) Long-term maintenance of hematopoiesis in irradiated mice by retrovirally transduced peripheral blood stem cells. Blood *89:* 1811–1817.

Drize, N.J., J.R. Keller, J.L. Chertkov (1996) Local clonal analysis of the hematopoietic system shows that multiple small short-living clones maintain life-long hematopoiesis in reconstituted mice. Blood *88:* 2927–2938.

Evans, M.J., M.H. Kaufman (1981) Establishment in culture of pluripotential cells from mouse embryos. Nature *292:* 154–156.

Fleming, W.H., E.J. Alpern, N. Uchida, K. Ikuta, G.J. Spangrude, I.L. Weissman (1993) Functional heterogeneity is associated with the cell cycle status of murine hematopoietic stem cells. J Cell Biol *122:* 897–902.

Francis, K., R. Ramakrishna, W. Holloway, B.O. Palsson (1998) Two new pseudopod morphologies displayed by the human hematopoietic KG1a progenitor cell line and by primary human CD34(+) cells. Blood *92:* 3616–3623.

Frimberger, A.E., C.I. McAuliffe, K.A. Werme, R.A. Tuft, K.E: Fogarty, B.O. Benoit, M.S. Dooner, P.J. Quesenberry (2001) The fleet feet of haematopoietic stem cells: rapid motility, interaction and proteopodia. Br J Haematol *112:* 644–654.

Geiger, H., S. Sick, C. Bonifer, A.M. Muller (1998) Globin gene expression is reprogrammed in chimeras generated by injecting adult hematopoietic stem cells into mouse blastocysts. Cell *93:* 1055–1065.

Gerike, T.G., U. Paulus, C.S. Potten, M. Loeffler (1998) A dynamic model of proliferation and differentiation in the intestinal crypt based on a hypothetical intraepithelial growth factor. Cell Prolif *31:* 93–110.

Goodell, M.A. (1999) CD34(+) or CD34(–): Does it really matter? Blood *94:* 2545–2547.

Goodell, M.A., K.A. Jackson, S.M. Majka, T. Mi, H. Wang, J. Pocius, C.J. Hartley, M.W. Majesky, M.L. Entman, L.H. Michael, K.K. Hirschi (2001) Stem cell plasticity in muscle and bone marrow. Ann N Y Acad Sci *938:* 208–218; discussion 218–220.

Gordon, M.Y., J.L. Lewis, S.B. Marley, F.H. Grand, J.M. Goldman (1997) Stromal cells negatively regulate primitive haemopoietic progenitor cell activation via a phosphatidylinositol-anchored cell adhesion/signalling mechanism. Br J Haematol *96:* 647–653.

Grossman, Z. (1986) The stem cell concept revisited: self-renewal capacity is a dynamic property of hemopoietic cells. Leuk Res *10:* 937–950.

Habibian, H.K., S.O. Peters, C.C. Hsieh, J. Wuu, K. Vergilis, C.I. Grimaldi, J. Reilly, J.E. Carlson, A.E. Frimberger, F.M. Stewart, P.J. Quesenberry (1998) The fluctuating phenotype of the lymphohematopoietic stem cell with cell cycle transit. J Exp Med *188:* 393–398.

Hao, Q.L., F.T. Thiemann, D. Petersen, E.M. Smogorzewska, G.M. Crooks (1996) Extended long-term culture reveals a highly quiescent and primitive human hematopoietic progenitor population. Blood *88:* 3306–3313.

Jordan, C.T., I.R. Lemischka (1990) Clonal and systemic analysis of long-term hematopoiesis in the mouse. Genes Dev *4:* 220–232.

Kapur, R., R. Cooper, L. Zhang, D.A. Williams (2001) Cross-talk between alpha(4)beta(1)/alpha(5)beta(1) and c-Kit results in opposing effect on growth and survival of hematopoietic cells via the activation of focal adhesion kinase, mitogen-activated protein kinase, and Akt signaling pathways. Blood *97:* 1975–1981.

Keller, G., H.R. Snodgrass (1999) Human embryonic stem cells: the future is now news. Nat Med *5:* 151–152.

Lemischka, I.R. (1997) Microenvironmental regulation of hematopoietic stem cells. Stem Cells *15 Suppl 1:* 63–68.

Loeffler, M., A. Birke, D. Winton, C. Potten (1993) Somatic mutation, monoclonality and stochastic models of stem cell organization in the intestinal crypt. J Theor Biol *160:* 471–491.

Loeffler, M., T. Bratke, U. Paulus, Y.Q. Li, C.S. Potten (1997) Clonality and life cycles of intestinal crypts explained by a state dependent stochastic model of epithelial stem cell organization. J Theor Biol *186:* 41–54.

Loeffler, M., B. Bungart, H. Goris, S. Schmitz, W. Nijhof (1989) Hemopoiesis during thiamphenicol treatment. II. A theoretical analysis shows consistency of new data with a previously hypothesized model of stem cell regulation. Exp Hematol *17:* 962–967.

Loeffler, M., B. Grossmann (1991) A stochastic branching model with formation of subunits applied to the growth of intestinal crypts. J Theor Biol *150:* 175–191.

Loeffler, M., C.S. Potten (1997) Stem cells and cellular pedigrees – a conceptual introduction; in Potten, C. S. (ed): Stem Cells. Cambridge, Academic Press, pp 1–27.

Loeffler, M., C.S. Potten, H.E. Wichmann (1987) Epidermal cell proliferation. II. A comprehensive mathematical model of cell proliferation and migration in the basal layer predicts some unusual properties of epidermal stem cells. Virchows Arch B Cell Pathol Incl Mol Pathol *53:* 286–300.

Loeffler, M., H.E. Wichmann (1980) A comprehensive mathematical model of stem cell proliferation which reproduces most of the published experimental results. Cell Tissue Kinet *13:* 543–561.

Lord, B.I. (1997) Biology of the haemopoietic stem cell; in Potten, C. S. (ed): Stem Cells. Cambridge, Academic Press, pp 401–422.

Lord, B.I., E.G. Wright (1984) Spatial organisation of CFU-S proliferation regulators in the mouse femur. Leuk Res *8:* 1073–1083.
Mackey, M.C., L. Glass (1977) Oscillation and chaos in physiological control systems. Science *197:* 287–289.
Meineke, F.A., C.S. Potten, M. Loeffler (2001) Cell migration and organization in the intestinal crypt using a lattice-free model. Cell Prolif *34*: 253–266.
Miller, S.J., R.M. Lavker, T.T. Sun (1997) Keratinocyte stem cells of cornea, skin and hair follicles; in Potten, C.S. (ed): Stem Cells. Cambridge, Academic Press, pp 331–362.
Muller-Sieburg, C.E., E. Deryugina (1995) The stromal cells' guide to the stem cell universe. Stem Cells *13:* 477–486.
Neben, S., D. Donaldson, C. Sieff, P. Mauch, D. Bodine, J. Ferrara, J. Yetz-Aldape, K. Turner (1994) Synergistic effects of interleukin-11 with other growth factors on the expansion of murine hematopoietic progenitors and maintenance of stem cells in liquid culture. Exp Hematol *22:* 353–359.
Ogawa, M., T.R. Mosmann (1985) Two stochastic models for stem cell differentiation. Leukemia: Recent Advances in Biology and Treatment 391–397.
Oostendorp, R.A., P. Dormer (1997) VLA-4-mediated interactions between normal human hematopoietic progenitors and stromal cells. Leuk Lymphoma *24:* 423–435.
Park, H.S., R.A. Goodlad, N.A. Wright (1995) Crypt fission in the small intestine and colon. A mechanism for the emergence of G6PD locus-mutated crypts after treatment with mutagens. Am J Pathol *147:* 1416–1427.
Paulus, U., M. Loeffler, J. Zeidler, G. Owen, C.S. Potten (1993) The differentiation and lineage development of goblet cells in the murine small intestinal crypt: experimental and modelling studies. J Cell Sci *106:* 473–483.
Paulus, U., C.S. Potten, M. Loeffler (1992) A model of the control of cellular regeneration in the intestinal crypt after perturbation based solely on local stem cell regulation. Cell Prolif *25*: 559–578.
Phillips, R.L., R.E. Ernst, B. Brunk, N. Ivanova, M.A. Mahan, J.K. Deanehan, K.A. Moore, Overton, G.C., I.R. Lemischka (2000) The genetic program of hematopoietic stem cells. Science *288:* 1635–1640.
Potten, C.S. (1991) Regeneration in epithelial proliferative units as exemplified by small intestinal crypts. Ciba Found Symp *160:* 54–71; discussion 71–76.
Potten, C.S. (1998) Stem cells in gastrointestinal epithelium: numbers, characteristics and death. Phil Trans R Soc Lond B Biol Sci *353:* 821–830.
Potten, C.S., M. Loeffler (1987a) A comprehensive model of the crypts of the small intestine of the mouse provides insight into the mechanisms of cell migration and the proliferation hierarchy. J Theor Biol *127:* 381–391.
Potten, C.S., M. Loeffler (1987b) Epidermal cell proliferation. I. Changes with time in the proportion of isolated, paired and clustered labelled cells in sheets of murine epidermis. Virchows Arch B Cell Pathol Incl Mol Pathol *53:* 279–285.
Potten, C.S., M. Loeffler (1990) Stem cells: attributes, cycles, spirals, pitfalls and uncertainties. Lessons for and from the crypt. Development *110:* 1001–1020.
Potten, C.S., H.E. Wichmann, M. Loeffler, K. Dobek, D. Major (1982) Evidence for discrete cell kinetic subpopulations in mouse epidermis based on mathematical analysis. Cell Tissue Kinet *15:* 305–329.
Quesenberry, P.J., P. Becker, F.M. Stewart (1998) Phenotype of the engrafting stem cell in mice. Stem Cells *16:* 33–35.
Rebel V.I., C.L. Miller, C.J. Eaves, P.M. Lansdorp (1996) The repopulation potential of fetal liver hematopoietic stem cells in mice exceeds that of their liver adult bone marrow counterparts. Blood *87:* 3500–3507.
Roberts, S.A., J.H. Hendry, C.S. Potten (1995) Deduction of the clonogen content of intestinal crypts: a direct comparison of two-dose and multiple-dose methodologies. Radiat Res *141:* 303–308.
Robertson, E.J. (1987) Embryo-derived stem cell lines; in Hames, D.R.A.B.D. (ed): Teratocarcinomas and embryonic stem cells: A practical approach. Oxford, IRL Press.
Roeder, I., M. Loeffler (2002) A novel dynamic model of hematopoietic stem cell organization based on the concept of within-tissue plasticity. Exp Hematol, in press.
Rosendaal, M., A. Mayen, A.D. Koning, T. Dunina-Barkovskaya, T. Krenacs, R. Ploemacher (1997) Does transmembrane communication through gap junctions enable stem cells to overcome stromal inhibition? Leukemia *11:* 1281–1289.
Roy, V., C.M. Verfaillie (1999) Expression and function of cell adhesion molecules on fetal liver, cord blood and bone marrow hematopoietic progenitors: implications for anatomical localization and developmental stage specific regulation of hematopoiesis. Exp Hematol *27:* 302–312.
Sato, T., J.H. Laver, M. Ogawa (1999) Reversible expression of CD34 by murine hematopoietic stem cells I. Blood *94:* 2548–2554.
Schofield, R. (1978) The relationship between the spleen colony-forming cell and the haemopoietic stem cell. Blood Cells *4:* 7–25.
Seale, P., M.A. Rudnicki (2000) A new look at the origin, function, and 'stem-cell' status of muscle satellite cells. Dev Biol *218:* 115–124.
Steel, G.G. (1977) Growth kinetics of tumors. Oxford, Clarendon Press.
Tajima, F., T. Sato, J.H. Laver, M. Ogawa (2000) CD34 expression by murine hematopoietic stem cells mobilized by granulocyte colony-stimulating factor. Blood *96:* 1989–1993.
Till, J.E., E.A. McCulloch, L. Siminovitch (1964) A stochastic model of stem cell proliferation, based on the growth of spleen colony-forming cells. Proc Natl Acad Sci USA *51:* 29–36.
Totafurno, J., M. Bjerknes, H. Cheng (1987) The crypt cycle. Crypt and villus production in the adult intestinal epithelium. Biophys J *52:* 279–294.
Trentin, J.J. (1971) Determination of bone marrow stem cell differentiation by stromal hemopoietic inductive microenvironments (HIM). Am J Pathol *65:* 621–628.
Uchida, N., W.H. Fleming, E.J. Alpern, I.L. Weissman (1993) Heterogeneity of hematopoietic stem cells. Curr Opin Immunol *5:* 177–184.
Van Zant, G., J.J. Chen K. Scott-Micus (1991) Developmental potential of hematopoietic stem cells determined using retrovirally marked allophenic marrow. Blood *77:* 756–763.
Van Zant, G., K. Scott-Micus, B.P. Thompson, R.A. Fleischman, S. Perkins (1992) Stem cell quiescence/activation is reversible by serial transplantation and is independent of stromal cell genotype in mouse aggregation chimeras. Exp Hematol *20:* 470–475.
Vermeulen, M., F. Le Pesteur, M.C. Gagnerault, J.Y. Mary, F. Sainteny, F. Lepault (1998) Role of adhesion molecules in the homing and mobilization of murine hematopoietic stem and progenitor cells. Blood *92:* 894–900.
Vogel, H., H. Niewisch, G. Matioli (1968) The self renewal probability of hemopoietic stem cells. J Cell Physiol *72:* 221–228.
Vogel, H., H. Niewisch, G. Matioli (1969) Stochastic development of stem cells. J Theor Biol *22*: 249–270.
Wichmann, H.E., M. Loeffler (1985) Mathematical modeling of cell proliferation: Stem cell regulation in hemopoiesis. Boca Raton, CRC Press, Vol. *1,2*.
Wichmann, H.E., M. Loeffler, S. Schmitz (1988) A concept of hemopoietic regulation and its biomathematical realization. Blood Cells *14:* 411–429.
Wineman, J., K. Moore, I. Lemischka, C. Muller-Sieburg (1996) Functional heterogeneity of the hematopoietic microenvironment: rare stromal elements maintain long-term repopulating stem cells. Blood *87:* 4082–4090.
Winton, D.J., M.A. Blount, B.A. Ponder (1988) A clonal marker induced by mutation in mouse intestinal epithelium. Nature *333:* 463–466.
Winton, D.J., B.A. Ponder (1990) Stem-cell organization in mouse small intestine. Proc R Soc Lond B Biol Sci *241:* 13–18.
Wright, N., M. Alison (1984) The biology of epithelial cell populations. Oxford, Clarendon Press.
Zhang, P., G. Behre, J. Pan, A. Iwama, N. Wara-Aswapati, H.S. Radomska, P.E. Auron, D.G. Tenen, Z. Sun (1999) Negative cross-talk between hematopoietic regulators: GATA proteins repress PU.1. Proc Natl Acad Sci USA *96:* 8705–8710.

Cells
Tissues
Organs

Cells Tissues Organs 2002;171:27–37

Hematopoietic Stem Cells: Self-Renewing or Aging?

Gerald de Haan

Department of Stem Cell Biology, University of Groningen, Groningen, The Netherlands

Key Words
Stem cells · Aging · Hematopoiesis · Plasticity

Abstract
Stem cells are defined by their extensive self-renewal properties, and yet there is abundant evidence of erosion of stem cell functioning during aging. Whereas intracellular repair and protection mechanisms determine the lifespan of an individual cell, here an argument is made that somatic stem cells determine the lifespan of an entire tissue, and thus play a key role in the organismal aging process. Recently it has been shown that the developmental potential of somatic stem cells may be far greater than previously anticipated. Although the mechanisms regulating stem cell plasticity are far from clear, it is challenging to discuss the potential relevance of these findings in the aging process.

Introduction

Recently, the scientific and indeed also the lay community has witnessed a surge of reports describing the unprecedented and unexpected biological properties of a class of cells that is collectively referred to as 'stem cells'. The current flurry of reports on stem cells and their potential clinical utility for a variety of human diseases merits a thorough analysis of concepts that are believed to rule stem cell functioning. Here we will focus on one such topic, namely stem cell aging. To discuss some of the recent findings properly and relate these to stem cell aging, it may be appropriate to reconsider carefully how stem cells are defined. Historically, a stem cell is defined as a functionally immature cell capable of generating large numbers of functionally mature, committed progeny, while simultaneously producing cells that are indistinguishable from the stem cell itself [Potten and Loeffler, 1990]. This latter process, generally referred to as self-renewal, ensures that the stem cell population itself is never exhausted. Stringent and loose definitions of stem cells have been coined in the literature. The most stringent definition probably would only consider the first generations of cells produced from a fertilized egg as true stem cells, whereas a very relaxed definition may even consider cells such as memory B-lymphocytes as stem cells, as both cell types meet the criteria of self-renewal and ability to produce massive numbers of mature cells. Obviously, the most explicit difference between cells formed after a fertilized egg first cleaves and a memory B-cell is the variety of distinct cell types each is able to produce. Cleavage divisions are uniquely confined to embryonic stem cells and are symmetric by definition. In contrast, more differentiated stem cells that undergo proliferation divisions have the option of both symmetric and asymmetric divi-

KARGER

Fax + 41 61 306 12 34
E-Mail karger@karger.ch
www.karger.com

1422–6405/02/1711–0027$18.50/0

Accessible online at:
www.karger.com/journals/cto

Gerald de Haan
Department of Stem Cell Biology
University of Groningen, A. Deusinglaan 1
NL–9713 AV Groningen (The Netherlands)
Tel. +31 50 3632722, Fax +31 50 3637477, E-Mail g.de.haan@med.rug.nl

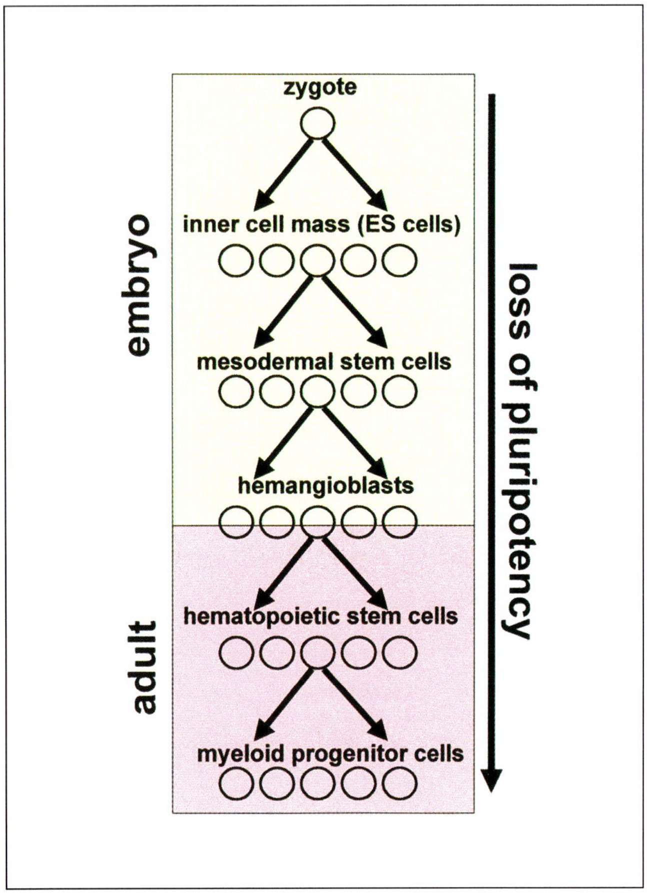

Fig. 1. A classical hierarchical stem cell model describing embryonic and adult (hematopoietic) cell production. During ontogeny, and continuing in adulthood, this hierarchical model implies that each cell division is necessarily associated with a loss of developmental potential. Furthermore, developmental properties that are lost cannot be regained.

sion. Whereas the committed daughter cells of a fertilized egg can both produce an entire organism (i.e. identical twins), the phenotype of memory B-cell offspring is very restricted and only consists of mature plasma cells. To differentiate between these two widely divergent models it is of practical use to introduce the concept of tissue-specific stem cells. Tissue-specific stem cells are capable of producing a variety of distinct cell types that collectively function to form a specific tissue. A well-known example includes hematopoietic stem cells, which are capable of producing massive numbers of lymphocytes, granulocytes, erythrocytes, monocytes, platelets, and additional derivatives hereof [Lajtha, 1975, 1979]. Also neural stem cells would qualify for this category, as they have been shown to be able to produce neurons, astroglia, and oligodendrocytes [Gage, 2000]. In fact, most tissues contain a population of such tissue-specific stem cells. To relate these separate stem cell populations developmentally, a concept has emerged of a hierarchy within a hierarchy (fig. 1). In this model, a small population of embryonic cells early in development in sequential steps produces a wide range of tissue-specific stem cells. The emergence of tissue-specific stem cells is carefully controlled, both in time and in space. For example, the first definitive hematopoietic stem cells in the mouse have been shown to arise early in development in a very specific region, called the Aorta-Gonad-Mesonephros (AGM), 10 days after conception [Müller et al., 1994; Medvinsky and Dzierzak, 1996]. During ontogeny, and continuing throughout adult life, tissue-specific stem cells generate large numbers of committed stem cells, that are capable of producing large numbers of fully differentiated, post-mitotic cells which carry out specific biological functions.

By definition, a hierarchy implies that fate decisions, either made instructively or stochastically, are of a top-to-bottom nature, and that developmental properties are gradually lost during differentiation. Furthermore, this concept favors the idea that properties once lost cannot be regained. Although this dogma had been challenged earlier, only recently the definitive experimental proof has been obtained by a large number of exciting findings that possibly imply that tissue-specific stem cells in fact may not have lost all developmental properties once possessed by their ancestors. Thus, adult somatic stem cells may prove to be far more plastic than previously imagined. In this review these new observations will be discussed in view of the role stem cells may play in the aging process.

Stem Cells and Aging

The mechanisms underlying the biological changes that occur during mammalian aging are still very poorly understood. Before Hayflick published his seminal paper on the limited replicative lifespan of mammalian cells growing in vitro [Hayflick and Moorhead, 1961; Hayflick, 1965], it was widely believed that organismal aging (why do we have a limited lifespan?) was unrelated to cellular aging (why do individual cells stop growing in culture after a defined period of time?). Our understanding of the molecular pathways involved in the induction of replicative senescence is still far from complete, but it is clear that the rate of organismal aging is the outcome of a complicated multifactorial machinery affecting the malfunctioning of individual cells. These individual cells, both in vitro and in vivo, are during their lifespan continuously exposed to damaged proteins and DNA molecules. Damage may occur randomly during DNA replication, but can be greatly increased by exogenous toxic compounds [Dolle et al., 2000; Vijg, 2000]. The significance of this kind of damage encountered by mammalian cells is most revealingly demonstrated in cells lacking key enzymes required to repair DNA damage, or degrade malfunctioning proteins. Mice, and often also humans, deficient in DNA repair genes such as *Ercc1*, *Csb*, and *Tfiih* show a very profound phenotype, which may be summarized collectively as accelerated aging [de Boer et al., 1998; van der Horst et al., 1997; Weeda et al., 1997]. Retardation of the aging process, in contrast, has been described in various animal models ranging from mouse to fly and worm, and so far has been shown to be closely associated with a reduction in levels of intracellular reactive oxygen species [Lin et al., 1998; Migliaccio et al., 1999; Melov et al., 2000]. Reactive oxygen species are highly toxic and have been shown to be able to produce both protein and DNA damage [Sohal and Weindruch, 1996]. If intracellular repair mechanisms fail to operate, or function at a less than optimal threshold, individual cells die rapidly by apoptosis. The limited lifespan of adult somatic cells (for example, only a few hours for neutrophilic granulocytes to many months for erythrocytes and even years for some lymphocyte subsets) implies that in order for any tissue to maintain its optimal physiological function, continuous replenishing of aged cells has to occur. It is here that stem cells come into play, since they form the crucial pool from which massive numbers of new cells can be derived. Thus, it seems likely that although the deleterious consequences of aging directly are targeted at the most widely abundant, post-mitotic, and fully matured cells present in all tissues of the organism, the price associated with the aging process in the long term is paid by tissue-specific stem cells. This concept may make the aging process substantially more complicated in stem cell-containing mammals compared to lower organisms. It may therefore be of relevance to note that in flies and worms, organisms that are essentially entirely post-mitotic and thus devoid of stem cell populations, a number of genes have been identified that affect organismal lifespan [Jazwinski, 2000; Kenyon et al., 1993; Kimura et al., 1997; Lin et al., 1997]; this has remained elusive in stem cell-containing mammals such as mice and human.

Thus, a challenging contradiction emerges: if stem cells were truly self-renewing and inexhaustible, as they should be by definition, they would indefinitely sustain the delivery of new cells to the various tissues, and aging would not occur. However, as deterioration of function can readily be demonstrated in almost any somatic tissue during aging, a conclusion may be that stem cells in fact are not able to self-renew forever. At a single cell level, aging and self-renewal are mutually exclusive. It is exactly because of this paradox that the concept of stem cell aging has been studied in considerable detail in the last three decades. Because of the unprecedented experimental model systems that have been, and still are, available for the exploration of hematopoietic stem cells, it is stem cell aging research in the field of hematology that has in particular advanced most dramatically.

Hematopoietic Stem Cells: Do They Age or Not?

Shortly after the first assays to quantify murine hematopoietic progenitor cells were established in the early 1960's, researchers tested the effect of age on stem cell functioning [Siminovitch et al., 1964]. An inherent problem associated with stem cell research, which potentially also confounds interpretation of many aging studies, is the fact that stem cells do not possess fixed properties but rather reveal their proliferative potential in certain experimental settings, and not others [Potten and Loeffler, 1990]. For the detection of hematopoietic stem cells, for example, assaying requires the transplantation of test cells into irradiated recipients followed by analysis of donor-derived colony-forming cells as they emerge in the bone marrow or spleen, or mature blood cells in the circulation [Harrison et al., 1988; Szilvassy et al., 1990]. Consequently, it has remained difficult to study the effect of age on hematopoietic stem cell functioning during normal, unperturbed steady state hematopoiesis.

Classic experiments testing the concept of indefinite self-renewal properties of hematopoietic stem cells used serial transplantation of bone marrow cells into lethally irradiated recipients. It was rapidly discovered that such consecutive transplantations could not be repeated very often; after 4–5 passages the repopulating ability of the bone marrow cells decreased dramatically, and transplanted mice succumbed as a consequence of marrow failure [Siminovitch et al., 1964; Ross et al., 1982]. Although these results were in concordance with the notion of a limited self-renewal capacity of stem cells (and thus seemed to reinforce the concept of replicative senescence, which had been postulated by Hayflick during the same period [Hayflick and Moorhead, 1961; Hayflick, 1965] alternative explanations for the apparent stem cell deficiency were coined. For example, it could not be excluded that the repopulation defect was initiated exactly by the highly artificial process of using cell transplantation to induce proliferative stress [Harrison and Astle, 1982]. Potentially, removal of stem cells from their own bone marrow microenvironment (in which they are normally firmly anchored by a complex set of adhesion molecules) as such might result in a loss of stem cells (or loss of stem cell function).

In addition, loss of repopulating ability of serially transplanted bone marrow cells potentially could also be explained by stem cell dilution. Since only a fraction of the total stem cell population is transplanted in this type of experiment (typically less than 1%), these limited numbers of stem cells are being dispersed throughout the entire recipient. Although the transplanted stem cells may undergo extensive self-renewal, complete restoration of stem cell numbers to pre-transplant values may not be reached. Subsequently, if bone marrow cells are obtained from these reconstituted mice and transplanted into a second cohort of recipients, the frequency of stem cells within this cell suspension will be lower than normal. It follows that each consecutive transplant contains fewer and fewer stem cells, although the self-renewal capacity per stem cell may have remained unaltered [Iscove and Nawa, 1997].

Using serial transplantation experiments it has been shown in C57BL/6 mice that stem cells can function to produce erythrocytes for time periods surpassing the lifespan of this strain of mice [Harrison, 1972]. These data clearly demonstrate the immense proliferative capacity of hematopoietic stem cells and originally were taken to indicate that organismal lifespan was not limited by cellular senescence. However, using the same arguments as mentioned earlier, care must be taken in extrapolating conclusions from these experimental data to stem cell functioning in *normal* hematopoiesis. Moreover, as will be discussed later, normal and perturbed stem cell functioning varies widely from strain to strain in mice, and the C57BL/6 strain may have exceptional aging properties, which prohibit drawing general conclusions from data obtained with this strain only.

The most convincing data demonstrating stem cell self-renewal directly, have been obtained in studies in which it was possible to track the proliferative fate of single stem cells. To this end two approaches have been applied. Several studies have used retroviral marking of unfractionated bone marrow cells to tag an individual stem cell, whose clonal progeny could be detected by analysis of the presence of unique retroviral integration sites [Capel et al., 1989, 1990; Jordan and Lemischka, 1990; Keller and Snodgrass, 1990]. These studies have shown that a single stem cell is capable of and sufficient to restore the entire hematopoietic system in lethally irradiated recipients. Earlier, it had been calculated that one mouse was able to repopulate 10^{15} recipient mice [Brian Lord, pers. comm.]. An alternative approach, documenting the same repopulating potential of a single stem cell, entailed the purification of primitive stem cells by multi-color flowcytometry, and the subsequent transplantation of one such cell (in conjunction with a low dose of unfractionated, genetically tagged, competitor bone marrow) in lethally irradiated recipients [Osawa et al., 1996]. Secondary transplantations have been performed using both approaches, and collectively these studies have proven that a single hematopoietic stem cell can divide and transfer long-term repopulating ability to its progeny.

All these data appeared to suggest that hematopoietic stem cells, either as a population or at the level of individual cells, do not show any marks of deteriorating function during aging. However, in conjunction with these data that fail to document an aging effect in stem cells, a large and accumulating body of evidence in fact does suggest that hematopoietic stem cells are not exempt from the age-related corrosion of potential that confronts almost any somatic cell. For example, it has been shown repeatedly, in a variety of experimental conditions, that hematopoietic stem cells isolated from the murine fetal liver have a marked increase in their proliferative potential compared with their adult counterparts [Micklem et al., 1972; Rebel et al., 1996a, b; Harrison et al., 1997]. In addition, the rate of cell cycling of hematopoietic cells declines gradually with age [Tejero et al., 1989; Kamminga et al., 2000]. Furthermore, during adult lifespan the number of stem cells changes as a function of age [Morrison et al., 1996; de Haan and Van Zant, 1999a; Chen et al., 1999, 2000]. These age-related changes have been shown to be highly strain-dependent. Stem cell numbers in C57BL/6 mice, that are historically most commonly used in hematopoietic stem cell research, increase significantly during aging. This effect has been documented using in vivo competitive repopulation assays [Harrison et al., 1989], in vitro cobblestone area forming cell formation [de Haan et al., 1997], and flowcytometric assays to quantify the most primitive cell subsets [Morrison et al., 1996]. However, using similar assays, stem cell numbers in all other strains of mice tested so far have been shown to drop significantly during aging [de Haan and Van Zant, 1999a; Chen et al., 1999, 2000].

Obviously, although it is relevant to quantify the available number of stem cells, it may be more important to determine the number of stem cells that at a given time point actively participates in the formation of new blood cells. It is currently widely believed that normal unperturbed hematopoiesis in young animals is a pluriclonal process (i.e. blood cell production is sustained by many stem cells simultaneously [Pietrzyk et al., 1985; Bradford et al., 1997; Cheshier et al., 1999], but shortly after stem cell transplantation it may very well be oligo- or indeed monoclonal [Jordan and Lemischka, 1990; Drize et al., 1996]. In mouse studies it remains to be determined whether as a consequence of the declining stem cell pool during aging steady state hematopoiesis becomes a mono/oligoclonal process in old animals. Clearly, a widespread conversion from pluriclonality to monoclonality in tissues with high rates of cell turnover would be compatible with the increased incidence of malignant cell growth during aging.

The nature of hematopoietic stem cell assays makes it far more difficult to obtain estimates of stem cell frequency in humans directly, and consequently data on the effects of aging on absolute stem cell numbers in humans do not exist. However, it has been demonstrated in many studies that hematopoiesis in humans shifts from a pluriclonal to an oligoclonal state during aging. These data rely indiscriminately on quantification of the extent of X chromosome inactivation in peripheral blood leukocytes from females at various ages. The existence of random X chromosome inactivation during the earliest stages of female development has resulted in a unique tool to address the issue of stem cell aging in humans. Due to irreversible transcriptional silencing of one of the two X chromosomes, in each female cell for any heterozygous locus only a single allele will be transcribed [Lyon, 1971]. Since inactivation is normally a random process, half of the leukocytes circulating in the peripheral blood of young females will express a paternal allele while the other half will express the maternal allele. It has been shown in many studies that in a large proportion of elderly women a significant skewing from the expected 50:50 ratio has occurred [Prchal et al., 1996; Champion et al., 1997; Gale et al., 1997; Tonon et al., 1998]. Interestingly, similar findings have been reported in cats, several years after they have received a bone marrow transplant with a relatively small number of cells [Abkowitz et al., 1995, 1996]. These findings are also compatible with the presence of X-linked genes directly affecting hematopoietic cell turnover [Abkowitz et al., 1998; de Haan and Van Zant, 1999a; Christensen et al., 2000] (and in fact such genes have actually been cloned from paroxysmal nocturnal haemoglobinuria patients [Bessler et al., 1994]). However, the frequency at which skewed X chromosome inactivation patterns are found in normal, healthy females strongly suggests that it is at least partly a consequence of a declining stem cell pool [Vickers et al., 2001].

The fact that in most strains of mouse and in most humans stem cells decline both in number and in function during normal aging, casts serious doubt on the concept of indefinite stem cell self-renewal. Recent data suggest that at least part of the erosion of stem cell function may result from the rate at which hematopoietic stem cells proliferate.

Rate of Cell Turn-Over Determines the Pace of Aging

As mentioned earlier, a general finding in cells that have undergone many rounds of cell divisions is deterioration of cell function, ultimately culminating in senescence. Although this concept has been most conclusively shown for cells grown in vitro, there is no evidence that it is restricted exclusively to cells cultured outside the body. The mechanism by which cells lose functional and proliferative potential as a result of their replicative history remains unknown. However, several mechanisms have been implicated and it is likely that they operate collectively and in concert. Briefly, malfunctioning may result from damaged genomic (including telomeric) or mitochondrial DNA, or from damaged proteins. Several recent reviews have addressed these issues in detail [Sohal and Weindruch, 1996; Finch and Tanzi, 1997; Guarente et al., 1998; Johnson et al., 1999].

The notion that deterioration of hematopoietic stem cell functioning occurs as a function of the rate at which these cells cycle, has been demonstrated in a series of studies in which stem cells obtained from a mouse strain with a high cell proliferation rate (DBA/2) were competed against stem cells obtained from a mouse strain with a low proliferation rate (C57BL/6) [Van Zant et al., 1990, 1991; Kamminga et al., 2000]. In these in vivo competition assays peripheral blood leukocytes were initially derived predominantly from rapidly cycling DBA/2 stem cells. However, during aging the contribution of these rapidly cycling stem cells to the peripheral leukocyte pool slowly decreased, whereas concomitantly the proportion of leukocytes from the slowly cycling C57BL/6 stem cells proportionally increased [Van Zant et al., 1990, 1991; Kamminga et al., 2000]. In keeping with the data described earlier, one explanation of this intriguing finding is that slowly cycling stem cells retain proliferative function for prolonged periods of time, whereas rapidly cycling stem cells burn their proliferative potential at higher rates. Interestingly, the DBA/2 strain of mice has a shorter lifespan than C57BL/6 mice [Russel, 1966; Gelman et al., 1988]. In an effort to test whether organismal lifespan and the rate of stem cell proliferation are indeed causally related, we have assessed the rate of cell proliferation in recombinant inbred BXD mice. Recombinant inbred BXD mice have been derived through repeated brother-sister matings, starting with C57BL/6 and DBA/2 mice as founders [Taylor et al., 1999]. Each BXD strain (of which 33 currently exist) carries a unique mosaic of C57BL/6 and DBA/2 genomic segments. Phenotyping the full panel of BXD strains for cell cycling rates, and subsequently correlating this trait with the known distribution of genotypes by quantitative trait analysis, has allowed the identification of genomic loci that were associated with variation in cell cycling rates [de Haan and Van Zant, 1999b]. A great advantage of recombinant inbred mice is that these strains are genetically stable, i.e. they are fully homozygous, and have been available to the research community for more than 15 years. In fact, careful longevity studies have been performed using these mice [Gelman et al., 1988]. Although separated in time more than 10 years, these studies have allowed us to correlate stem cell cycling and mean mouse lifespan directly [de Haan and Van Zant, 1999b]. Strikingly, a strong inverse correlation between hematopoietic cell cycling and organismal lifespan was observed. Comparison of genomic regions associated with each trait resulted in the identification of two loci, mapping to chromosome 7 and 11 that were involved in both cell cycling and lifespan. Although the underlying genes have yet to be identified, these data strongly argue for a direct connection between in vivo cell proliferation rates and lifespan. We have recently generated C57BL/6 congenic mice that carry a segment of DBA/2 chromosome 11 containing the cell cycle/lifespan locus. Preliminary data have documented a significant stem cell defect in these mice [GdH, April 2001].

Additional evidence for the direct involvement of cell turnover rate and stem cell depletion is supplied by a careful analysis of the hematopoietic system of mice deficient in the cyclin-dependent kinase inhibitor p21 [Cheng et al., 2000]. Although p21 is apparently dispensable for normal blood cell production, stem cells from p21–/– mice are highly proliferative, and are more easily depleted after serial bone marrow transplantation, or after chemotherapy. As such, p21 plays an important role in preserving the primitive stem cell population in conditions of high replicative stress.

Collectively, the data reviewed here suggest that hematopoietic stem cells indeed lose proliferative potential during aging. In fact, such loss of potential has been widely believed to be the necessary consequence of normal development. Beginning only a very few divisions after the egg has been fertilized by sperm, embryonic cells become increasingly committed in their proliferative potential. In humans this is most evidently documented by the absence of monozygotic quadruples. Referring to the concept of stem cell hierarchies introduced in figure 1, this loss of potential was, for a long time, believed to be irreversible: cellular rejuvenation was not considered to be an option. However, experimental data that has been published recently is challenging this dogma. In these studies that will be briefly reviewed in the next paragraph somatic adult stem cells harvested from a variety of tissues (but most notably from the bone marrow) have been shown to be far more potent than once believed. How do these data fit into the concept of stem cell aging?

Stem Cell Plasticity and Its Significance in Aging: Back to the Future?

Dolly, the sheep that was born seemingly normal and healthy from the transfer of a nucleus from an adult cell into an enucleated egg, demonstrated that somatic cellular aging does not necessarily and irreversibly alter the genetic composition of a cell [Wilmut et al., 1997]. If an aged nucleus is heterochronically transplanted into an enucleated egg its full developmental program can be reinitiated. However, since this first report it has become increasingly clear that the developmental program of most cloned animals is not normal, the many animals with abnormal phenotypes possibly resulting from failure to erase adult DNA methylation patterns [Jaenisch and Wilmut, 2001]. In addition, the observation that mammalian cloning has a very low success rate [Wakayama et al., 1998; Kato et al., 1998; Baguisi et al., 1999; Polejaeva et al., 2000] may in part be due to the fact that the majority of adult somatic cells in fact does not possess a fully reprogrammable genome, which might be consequence of aging.

Nevertheless, the various cloning experiments have proven that the microenvironment in which a nucleus finds itself can determine the developmental program that will ensue. By the same corollary, a stem cell may only reveal its full developmental program when it is positioned in the proper environment. A hematopoietic stem cell, residing in the bone marrow, will normally only pro-

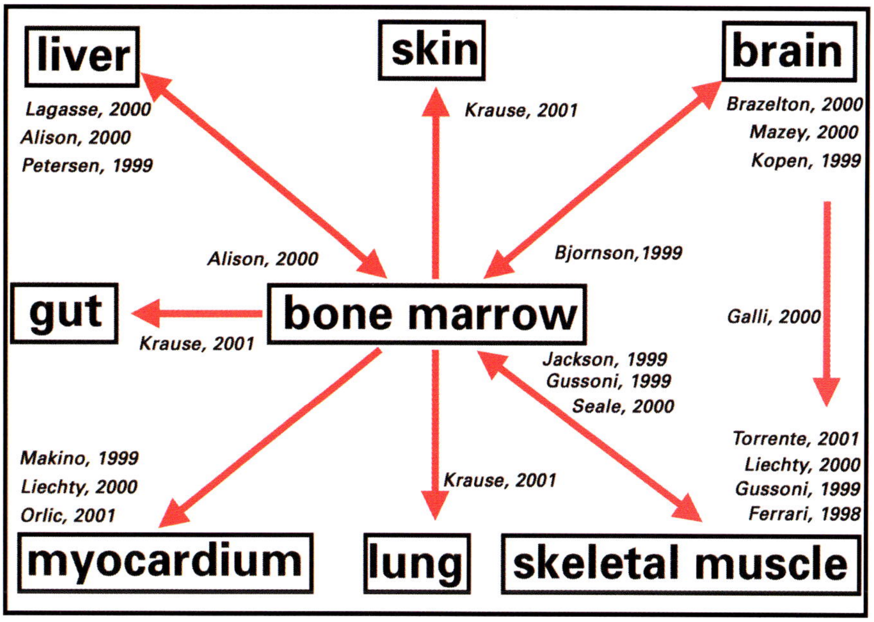

Fig. 2. Stem cell plasticity. It has been shown that the bone marrow contains hematopoietic and/or mesenchymal cell populations that are able to produce such diverse cell types as hepatocytes, neuronal cells, cardiomyocytes and skeletal muscle cells upon in vivo heterologous transplantation or in vitro differentiation. References indicate the original experimental studies.

duce mature blood cells. However, it has recently been shown that if hematopoietic stem cells are transplanted into properly conditioned recipient mice they are capable of producing hepatic, neural, skeletal, lung, gastrointestinal tract, skin, and cardiac myogenic cells [Ferrari et al., 1998; Makino et al., 1999; Gussoni et al., 1999; Kopen et al., 1999; Petersen et al., 1999; Alison et al., 2000; Brazelton et al., 2000; Lagasse et al., 2000; Liechty et al., 2000; Mezey et al., 2000; Orlic et al., 2001; Torrente et al., 2001, Krause et al., 2001] (fig. 2). Similar findings have been reported for stem cells derived from other tissues than bone marrow. For example, liver cells have been shown to contribute to blood cell formation [Alison et al., 2000]. Also, neural cells can produce blood cells when transplanted into lethally irradiated recipients [Bjornson et al., 1999], skeletal muscle [Galli et al., 2000] and may contribute to almost any tissue if injected into blastocysts [Clarke et al., 2000]. A cell population in skeletal muscle has been shown to have extensive hematopoietic potential [Gussoni et al., 1999; Jackson et al., 1999; Seale et al., 2000].

At present it is far from clear how to interpret these data. Assuming that it will be possible to reproduce these surprising data by independent research groups, at least four distinct stem cell models can be proposed (fig. 3). The most challenging concept argues that the genetic program of tissue-restricted stem cells is plastic and can be reprogrammed by the environment in which the cells reside (model 1, fig. 3). Immediately after lodging of tissue-specific stem cells into a new environment extracellular cues would initiate novel genes to be transcribed, adjusting the transplanted cell to its new milieu. Several studies have indeed suggested or implied that very shortly after transplantation the gene expression profile and function of transplanted stem cells may alter dramatically [Clarke et al., 2000; Orlic et al., 2001]. Stem cell plasticity would therefore be conceptually similar to transfer of an adult nucleus into an enucleated egg. An alternative, although related explanation, would suggest that tissue-specific stem cells have the capacity to dedifferentiate towards a more primitive state, thus gaining developmental potential (model 2, fig. 3). This hypothesis would predict that it might take considerable time before the newly formed offspring of heterologously transplanted stem cells would contribute to tissue formation in their new location. Again, some evidence is in accordance with this concept, as it takes many weeks before intravenously transplanted neural stem cells contribute to blood formation [Bjornson et al., 1999]. Thirdly, it may be possible that stem cells that are restricted in the types of cells they can produce are present in, at first sight, 'inappropriate' tissues (model 3, fig. 3). For example, stem cells with exclusively hematopoietic potential may reside in skeletal muscle, or in fact in any other tissue. Finally, it cannot be excluded that embryonic stem cell-like cells that maintain a wide developmental potential may survive in the adult

Fig. 3. New stem cell models that may explain apparent stem cell plasticity. If tissue A and tissue B both function normally (left panels), tissue-specific stem cells A and B contribute to their respective lineages exclusively. However, if tissue B functions improperly and cells from tissue A are transplanted into tissue B, B-type cells may be produced by various mechanisms (right panels). Model 1 considers the possibility that an A-type stem cell only behaves A-like because it is located in tissue A. Upon transplantation to tissue B, it receives novel cues from the B-type environment which directly reprograms the transcriptional activity from an A-type to a B-type, and converts the stem cell identity from A to B. In the second model an A-type stem cell upon transplantation from tissue A to tissue B actually dedifferentiates and thereby gains B-type developmental potential. It converts to a bipotential AB-type stem cell before it is forced by the tissue B environment to exclusively produce B-type cells. The third model considers the option that tissue A normally contains type-B stem cells, that remain quiescent as they do not receive the appropriate signals from the tissue A environment. Upon transplantation to tissue B, however, these dormant B-type stem cells become active whereas type-A stem cells stop dividing. The fourth model takes into account the possibility that bipotent AB-type stem cells are present in tissue A. These multipotent (embryonic?) stem cells normally may function to produce committed A-type stem cells. However, after transplant these primitive stem cells produce B-type stem cells in their new B environment, and stop producing A-type stem cells. Whereas model 1 and 2 would be compatible with the notion of actual stem cell plasticity, model 3 and 4 are not.

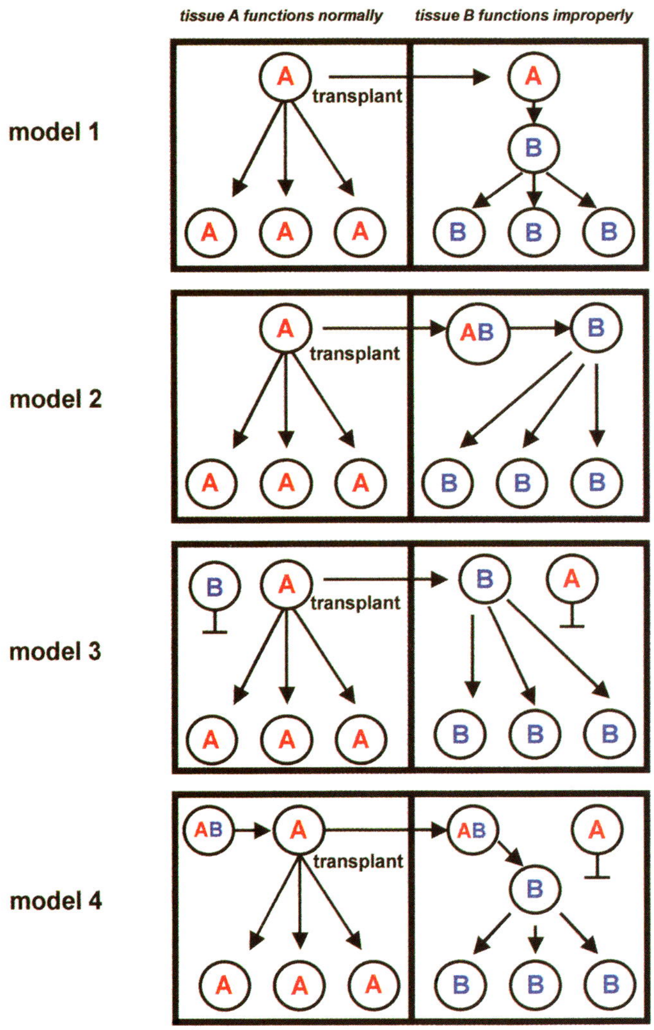

organism for very prolonged periods of time (at least far into adulthood, model 4, fig. 3). Such cells may never self-renew but remain quiescent for most of the life time of the individual. Importantly, these various models need not be mutually exclusive, but may be operating simultaneously. The definitive proof that all the various types of mature blood cells were all derived from a common ancestor required the demonstration of their clonal decent. Similarly, crucial experiments in the still infant field of stem cell plasticity will involve clonal analysis, either through retroviral marking of transplantable cells and tracing of clonal progeny in multiple tissues, or alternatively, demonstrating that multiple tissues can be (partly) reconstituted from one individual stem cell. This last goal may have been met in a recent study [Krause et al., 2001].

Obviously these experimental data are very exciting, and their potential clinical ramifications are imminent. However, from a fundamental point of view the most pertinent issue is whether they have any relevance for normal, in vivo, cell production. All data published so far have only made observations in highly artificial transplant settings. If neural stem cells can contribute to blood cell formation in lethally irradiated recipients, can and do they also participate in normal steady-state hematopoiesis? If hematopoietic stem cells can form hepatocytes after transplant in animals suffering from severe liver disease, can and will they also do this in unperturbed animals? The answer remains unknown so far, but it is an easy prediction that many studies will appear in the very near future addressing exactly this issue.

From an aging point of view data on stem cell plasticity are particularly intriguing. If it turns out that in unperturbed organisms the developmental potential of somatic stem cells indeed exceeds by far the confinement of the tissue in which they reside, one may speculate that they play a crucial role in the organismal aging process. As a result of both genetic and environmental influences, certain tissues will, with age, be more impaired in functioning than others. Those organs that cease to function first will ultimately limit the physiological integrity of an aging organism. It is a challenging speculation to propose that whereas DNA repair and protein protection mechanisms may extend the lifespan of an individual cell, stem cells may extend the lifespan of an organism. The extent of stem cell plasticity, by whatever mechanism it may be regulated, might thus interfere with the rate at which organismal aging occurs.

Acknowledgements

The work in the lab of GdH is supported by grants from the Royal Netherlands Academy of Arts and Sciences (KNAW), the Netherlands Organization for Scientific Research (NWO), and the Dutch Cancer Society.

I thank Dr. Christine Mummery, Dr. Piet Wierenga and Dr. Brian Lord for critically reading the manuscript and offering valuable suggestions for improvements.

References

Abkowitz, J.L., S.N. Catlin, P. Guttorp (1996) Evidence that hematopoiesis may be a stochastic process in vivo. Nat Med *2:* 190–197.

Abkowitz, J.L., M.T. Persik, G.H. Shelton, R.L. Ott, J.V. Kiklevich, S.N. Catlin, P. Guttorp (1995). Behavior of hematopoietic stem cells in a large animal. Proc Natl Acad Sci USA *92:* 2031–2035.

Abkowitz, J.L., M. Taboada, G.H. Shelton, S.N. Catlin, P. Guttorp, J.V. Kiklevich (1998) An X chromosome gene regulates hematopoietic stem cell kinetics. Proc Natl Acad Sci USA *95:* 3862–3866.

Alison, M.R., R. Poulsom, R. Jeffery, A.P. Dhillon, A. Quaglia, J. Jacob, M. Novelli, G. Prentice, J. Williamson, N.A. Wright (2000) Hepatocytes from non-hepatic adult stem cells. Nature *406:* 257.

Baguisi, A., E. Behboodi, D.T. Melican, J.S. Pollock, M.M. Destrempes, C. Cammuso, J.L. Williams, S.D. Nims, C.A. Porter, P. Midura, M.J. Palacios, S.L. Ayres, R.S. Denniston, M.L. Hayes, C.A. Ziomek, H.M. Meade, R.A. Godke, W.G. Gavin, E.W. Overstrom, Y. Echelard (1999) Production of goats by somatic cell nuclear transfer. Nat Biotechnol *17:* 456–61.

Bessler, M., P.J. Mason, P. Hillmen, T. Miyata, N. Yamada, J. Takeda, L. Luzzatto, T. Kinoshita (1994) Paroxysmal nocturnal haemoglobinuria (PNH) is caused by somatic mutations in the PIG-A gene. EMBO J *13:* 110–7.

Bjornson, C.R., R.L. Rietze, B.A. Reynolds, M.C. Magli, A.L. Vescovi (1999) Turning brain into blood: a hematopoietic fate adopted by adult neural stem cells in vivo [see comments]. Science *283:* 534–537.

Bradford, G.B., B. Williams, R. Rossi, I. Bertoncello (1997) Quiescence, cycling, and turnover in the primitive hematopoietic stem cell compartment. Exp Hematol *25:* 445–453.

Brazelton, T.R., F.M. Rossi, G.I. Keshet, H.M. Blau (2000) From marrow to brain: expression of neuronal phenotypes in adult mice. Science *290:* 1775–1779.

Capel, B., R. Hawley, L. Covarrubias, T. Hawley, B. Mintz (1989) Clonal contributions of small numbers of retrovirally marked hematopoietic stem cells engrafted in unirradiated neonatal W/Wv mice. Proc Natl Acad Sci USA *86:* 4564–4568.

Capel, B., R.G. Hawley, B. Mintz (1990) Long- and short-lived murine hematopoietic stem cell clones individually identified with retroviral integration markers. Blood *75:* 2267–70.

Champion, K.M., J.G. Gilbert, F.A. Asimakopoulos, S. Hinshelwood, A.R. Green (1997) Clonal haemopoiesis in normal elderly women: implications for the myeloproliferative disorders and myelodysplastic syndromes. Br J Haematol *97:* 920–926.

Chen, J., C.M. Astle, D.E. Harrison (1999) Development and aging of primitive hematopoietic stem cells in BALB/cBy mice. Exp Hematol *27:* 928–935.

Chen, J., C.M. Astle, D.E. Harrison (2000) Genetic regulation of primitive hematopoietic stem cell senescence. Exp Hematol *28:* 442–450.

Cheng, T., N. Rodrigues, H. Shen, Y. Yang, D. Dombkowski, M. Sykes, D.T. Scadden (2000) Hematopoietic stem cell quiescence maintained by p21cip1/waf1. Science *287:* 1804–1808.

Cheshier, S.H., S.J. Morrison, X. Liao, I.L. Weissman (1999) In vivo proliferation and cell cycle kinetics of long-term self-renewing hematopoietic stem cells. Proc Natl Acad Sci USA *96:* 3120–3125.

Christensen, K., M. Kristiansen, H. Hagen-Larsen, A. Skytthe, L. Bathum, B. Jeune, K. Andersen-Ranberg, J.W. Vaupel, K.H. Orstavik (2000) X-linked genetic factors regulate hematopoietic stem-cell kinetics in females. Blood *95:* 2449–2451.

Clarke, D.L., C.B. Johansson, J. Wilbertz, B. Veress, E. Nilsson, H. Karlstrom, U. Lendahl, J. Frisen (2000) Generalized potential of adult neural stem cells [see comments]. Science *288:* 1660–1663.

de Boer, J., J. de Wit, H. van Steeg, R.J. Berg, H. Morreau, P. Visser, A.R. Lehmann, M. Duran, J.H. Hoeijmakers, G. Weeda (1998) A mouse model for the basal transcription/DNA repair syndrome trichothiodystrophy. Mol Cell *1:* 981–990.

de Haan, G., W. Nijhof, G. Van Zant (1997) Mouse strain-dependent changes in frequency and proliferation of hematopoietic stem cells during aging: correlation between lifespan and cycling activity. Blood *89:* 1543–1550.

de Haan, G., G. Van Zant (1999a) Dynamic changes in mouse hematopoietic stem cell numbers during aging. Blood *93:* 3294–3301.

de Haan, G., G. Van Zant (1999b) Genetic analysis of hemopoietic cell cycling in mice suggests its involvement in organismal life span. Faseb J *13:* 707–713.

Dolle, M.E., W.K. Snyder, J.A. Gossen, P.H. Lohman, J. Vijg (2000) Distinct spectra of somatic mutations accumulated with age in mouse heart and small intestine. Proc Natl Acad Sci USA *97:* 8403–8408.

Drize, N.J., J.R. Keller, J.L. Chertkov (1996) Local clonal analysis of the hematopoietic system shows that multiple small short-living clones maintain life-long hematopoiesis in reconstituted mice. Blood *88:* 2927–2938.

Ferrari, G., G. Cusella-De Angelis, M. Coletta, E. Paolucci, A. Stornaiuolo, G. Cossu, F. Mavilio (1998) Muscle regeneration by bone marrow-derived myogenic progenitors [see comments] [published erratum appears in Science 1998 Aug 14;281(5379):923]. Science *279:* 1528–1530.

Finch, C.E., R.E. Tanzi (1997) Genetics of aging. Science *278:* 407–411.

Gage, F.H. (2000) Mammalian neural stem cells. Science *287:* 1433–1438.

Gale, R.E., A.K. Fielding, C.N. Harrison, D.C. Linch (1997) Acquired skewing of X-chromosome inactivation patterns in myeloid cells of the elderly suggests stochastic clonal loss with age. Br J Haematol *98:* 512–519.

Galli, R., U. Borello, A. Gritti, M.G. Minasi, C. Bjornson, M. Coletta, M. Mora, M.G. De Angelis, R. Fiocco, G. Cossu, A.L. Vescovi (2000) Skeletal myogenic potential of human and mouse neural stem cells. Nat Neurosci *3:* 986–991.

Gelman, R., A. Watson, R. Bronson, E. Yunis (1988) Murine chromosomal regions correlated with longevity. Genetics *118:* 693–704.

Guarente, L., G. Ruvkun, R. Amasino (1998) Aging, life span, and senescence. Proc Natl Acad Sci USA *95:* 11034–11036.

Gussoni, E., Y. Soneoka, C.D. Strickland, E.A. Buzney, M.K. Khan, A.F. Flint, L.M. Kunkel, R.C. Mulligan (1999) Dystrophin expression in the mdx mouse restored by stem cell transplantation. Nature *401:* 390–394.

Harrison, D.E. (1972) Normal function of transplanted mouse erythrocyte precursors for 21 months beyond donor life spans. Nature New Biol *237:* 220–222.

Harrison, D.E., C.M. Astle (1982) Loss of stem cell repopulating ability upon transplantation. Effects of donor age, cell number, and transplantation procedure. J Exp Med *156:* 1767–1779.

Harrison, D.E., C.M. Astle, C. Lerner (1988) Number and continuous proliferative pattern of transplanted primitive immunohematopoietic stem cells. Proc Natl Acad Sci USA *85:* 822–826.

Harrison, D.E., C.M. Astle, M. Stone (1989) Numbers and functions of transplantable primitive immunohematopoietic stem cells. Effects of age. J Immunol *142:* 3833–3840.

Harrison, D.E., R.K. Zhong, C.T. Jordan, I.R. Lemischka, C.M. Astle (1997) Relative to adult marrow, fetal liver repopulates nearly five times more effectively long-term than short-term. Exp Hematol *25:* 293–297.

Hayflick, L. (1965) The limited in vitro lifetime of human diploid cell strains. Exp Cell Res *37:* 614–636.

Hayflick, L., P. Moorhead (1961) The serial cultivation of human diploid cell strains. Exp Cell Res *25:* 585–621.

Iscove, N.N., K. Nawa (1997) Hematopoietic stem cells expand during serial transplantation in vivo without apparent exhaustion. Curr Biol *7:* 805–808.

Jackson, K.A., T. Mi, M.A. Goodell (1999) Hematopoietic potential of stem cells isolated from murine skeletal muscle [see comments]. Proc Natl Acad Sci USA *96:* 14482–14486.

Jaenisch, R., I. Wilmut (2001) Developmental biology. Don't clone humans! Science *291:* 2552.

Jazwinski, S.M. (2000) Aging and longevity genes. Acta Biochim Pol *47:* 269–279.

Johnson, F.B., D.A. Sinclair, L. Guarente (1999) Molecular biology of aging. Cell *96:* 291–302.

Jordan, C.T., I.R. Lemischka (1990) Clonal and systemic analysis of long-term hematopoiesis in the mouse. Genes Dev *4:* 220–232.

Kamminga, L.M., I. Akkerman, E. Weersing, A. Ausema, B. Dontje, G. Van Zant, G. de Haan (2000) Autonomous behavior of hematopoietic stem cells. Exp Hematol *28:* 1451–1459.

Kato, Y., T. Tani, Y. Sotomaru, K. Kurokawa, J. Kato, H. Doguchi, H. Yasue, Y. Tsunoda (1998) Eight calves cloned from somatic cells of a single adult. Science *282:* 2095–2098.

Keller, G., R. Snodgrass (1990) Life span of multipotential hematopoietic stem cells in vivo. J Exp Med *171:* 1407–1418.

Kenyon, C., J. Chang, E. Gensch, A. Rudner, R. Tabtiang (1993) A C. elegans mutant that lives twice as long as wild type. Nature *366:* 461–464.

Kimura, K.D., H.A. Tissenbaum, Y. Liu, G. Ruvkun (1997) daf-2, an insulin receptor-like gene that regulates longevity and diapause in Caenorhabditis elegans [see comments]. Science *277:* 942–946.

Kopen, G.C., D.J. Prockop, D.G. Phinney (1999) Marrow stromal cells migrate throughout forebrain and cerebellum, and they differentiate into astrocytes after injection into neonatal mouse brains. Proc Natl Acad Sci USA *96:* 10711–10716.

Krause, D., N.D. Theise, M.I. Collector, O. Henegariu, S. Hwang, R. Gardner, S. Neutzel, S.J. Sharkis (2001) Multi-organ, multi-lineage engraftment by a single bone marrow-derived stem cell. Cell *105:* 369–377.

Lagasse, E., H. Connors, M. Al-Dhalimy, M. Reitsma, M. Dohse, L. Osborne, X. Wang, M. Finegold, I.L. Weissman, M. Grompe (2000) Purified hematopoietic stem cells can differentiate into hepatocytes in vivo. Nat Med *6:* 1229–1234.

Lajtha L.G. (1975) Haematopoietic stem cells. Br J Haematol *29:* 529–535.

Lajtha L.G. (1979) Stem cell concepts. Differentiation *14:* 23–34.

Liechty, K.W., T.C. MacKenzie, A.F. Shaaban, A. Radu, A.M. Moseley, R. Deans, D.R. Marshak, A.W. Flake (2000) Human mesenchymal stem cells engraft and demonstrate site-specific differentiation after in utero transplantation in sheep. Nat Med *6:* 1282–1286.

Lin, K., J.B. Dorman, A. Rodan, C. Kenyon (1997) daf-16: an HNF-3/forkhead family member that can function to double the life-span of Caenorhabditis elegans. Science *278:* 1319–1322.

Lin, Y.J., L. Seroude, S. Benzer (1998) Extended life-span and stress resistance in the Drosophila mutant methuselah [see comments]. Science *282:* 943–946.

Lyon, M.F. (1971) Possible mechanisms of X chromosome inactivation. Nat New Biol *232:* 229–232.

Makino, S., K. Fukuda, S. Miyoshi, F. Konishi, H. Kodama, J. Pan, M. Sano, T. Takahashi, S. Hori, H. Abe, J. Hata, A. Umezawa, S. Ogawa (1999) Cardiomyocytes can be generated from marrow stromal cells in vitro. J Clin Invest *103:* 697–705.

Medvinsky, A., E. Dzierzak (1996) Definitive hematopoiesis is autonomously initiated by the AGM region. Cell *86:* 897–906.

Melov, S., J. Ravenscroft, S. Malik, M.S. Gill, D.W. Walker, P.E. Clayton, D.C. Wallace, B. Malfroy, S.R. Doctrow, G.J. Lithgow (2000) Extension of life-span with superoxide dismutase/catalase mimetics. Science *289:* 1567–1569.

Mezey, E., K.J. Chandross, G. Harta, R.A. Maki, S.R. McKercher (2000) Turning blood into brain: cells bearing neuronal antigens generated in vivo from bone marrow. Science *290:* 1779–1782.

Micklem, H.S., C.E. Ford, E.P. Evans, D.A. Ogden, D.S. Papworth (1972) Competitive in vivo proliferation of foetal and adult haematopoietic cells in lethally irradiated mice. J Cell Physiol *79:* 293–298.

Migliaccio, E., M. Giorgio, S. Mele, G. Pelicci, P. Reboldi, P.P. Pandolfi, L. Lanfrancone, P.G. Pelicci (1999) The p66shc adaptor protein controls oxidative stress response and life span in mammals. Nature *402:* 309–313.

Morrison, S.J., A.M. Wandycz, K. Akashi, A. Globerson, I.L. Weissman (1996) The aging of hematopoietic stem cells. Nat Med *2:* 1011–1016.

Müller, A.M., A. Medvinsky, J. Strouboulis, F. Grosveld, E. Dzierzak (1994) Development of hematopoietic stem cell activity in the mouse embryo. Immunity *1:* 291–301.

Orlic, D., J. Kajstura, S. Chimenti, I. Jakoniuk, S.M. Anderson, B. Li, J. Pickel, R. McKay, B. Nadal-Ginard, D.M. Bodine, A. Leri, P. Anversa (2001) Bone marrow cells regenerate infarcted myocardium. Nature *410:* 701–705.

Osawa, M., K. Hanada, H. Hamada, H. Nakauchi (1996) Long-term lymphohematopoietic reconstitution by a single CD34-low/negative hematopoietic stem cell. Science *273:* 242–245.

Petersen, B.E., W.C. Bowen, K.D. Patrene, W.M. Mars, A.K. Sullivan, N. Murase, S.S. Boggs, J.S. Greenberger, J.P. Goff (1999) Bone marrow as a potential source of hepatic oval cells. Science *284:* 1168–1170.

Pietrzyk, M.E., G.V. Priestley, N.S. Wolf (1985) Normal cycling patterns of hematopoietic stem cell subpopulations: an assay using long-term in vivo BrdU infusion. Blood *66:* 1460–1462.

Polejaeva, I.A., S.H. Chen, T.D. Vaught, R.L. Page, J. Mullins, S. Ball, Y. Dai, J. Boone, S. Walker, D.L. Ayares, A. Colman, K.H. Campbell (2000) Cloned pigs produced by nuclear transfer from adult somatic cells. Nature *407:* 86–90.

Potten, C.S., M. Loeffler (1990) Stem cells: attributes, cycles, spirals, pitfalls and uncertainties. Lessons for and from the crypt. Development *110:* 1001–1020.

Prchal, J.T., J.F. Prchal, M. Belickova, S. Chen, Y. Guan, G.L. Gartland, M.D. Cooper (1996) Clonal stability of blood cell lineages indicated by X-chromosomal transcriptional polymorphism. J Exp Med *183:* 561–567.

Rebel, V.I., C.L. Miller, C.J. Eaves, P.M. Lansdorp (1996a) The repopulation potential of fetal liver hematopoietic stem cells in mice exceeds that of their liver adult bone marrow counterparts. Blood *87:* 3500–3507.

Rebel, V.I., C.L. Miller, G.R. Thornbury, W.H. Dragowska, C.J. Eaves, P.M. Lansdorp (1996b) A comparison of long-term repopulating hematopoietic stem cells in fetal liver and adult bone marrow from the mouse. Exp Hematol *24:* 638–648.

Ross, E.A., N. Anderson, H.S. Micklem (1982) Serial depletion and regeneration of the murine hematopoietic system. Implications for hematopoietic organization and the study of cellular aging. J Exp Med *155:* 432–444.

Russel, E.S. (1966) Lifespan and aging patterns; in Green, E.L. (ed): Biology of the Laboratory Mouse., vol. 2. pp. 511–519. New York, McGraw-Hill.

Seale, P., L.A. Sabourin, A. Girgis-Gabardo, A. Mansouri, P. Gruss, M.A. Rudnicki (2000) Pax7 is required for the specification of myogenic satellite cells [In Process Citation]. Cell *102:* 777–786.

Siminovitch, L., J.E. Till, E.A. Mcculloch (1964) Decline in colony forming ability of marrow cells subjected to serial transplantation into irradiated mice. J Cell Comp Physiol *64:* 23–31.

Sohal, R.S., R. Weindruch (1996) Oxidative stress, caloric restriction, and aging. Science *273:* 59–63.

Szilvassy, S.J., R.K. Humphries, P.M. Lansdorp, A.C. Eaves, C.J. Eaves (1990) Quantitative assay for totipotent reconstituting hematopoietic stem cells by a competitive repopulation strategy. Proc Natl Acad Sci USA *87:* 8736–8740.

Taylor, B.A., C. Wnek, B.S. Kotlus, N. Roemer, T. MacTaggart, S.J. Phillips (1999) Genotyping new BXD recombinant inbred mouse strains and comparison of BXD and consensus maps. Mamm Genome *10:* 335–348.

Tejero, C., N.G. Testa, J.H. Hendry (1989) Decline in cycling of granulocyte-macrophage colony-forming cells with increasing age in mice. Exp Hematol *17:* 66–67.

Tonon, L., G. Bergamaschi, C. Dellavecchia, V. Rosti, C. Lucotti, L. Malabarba, A. Novella, E. Vercesi, F. Frassoni, M. Cazzola (1998) Unbalanced X-chromosome inactivation in haemopoietic cells from normal women. Br J Haematol *102:* 996–1003.

Torrente, Y., J.P. Tremblay, F. Pisati, M. Belicchi, B. Rossi, M. Sironi, F. Fortunato, M. El Fahime, M.G. D'Angelo, N.J. Caron, G. Constantin, D. Paulin, G. Scarlato, N. Bresolin (2001) Intraarterial Injection of Muscle-derived CD34(+)Sca-1(+) Stem cells restores dystrophin in mdx Mice. J Cell Biol *152:* 335–348.

van der Horst, G.T., H. van Steeg, R.J. Berg, A.J. van Gool, J. de Wit, G. Weeda, H. Morreau, R.B. Beems, C.F. van Kreijl, F.R. de Gruijl, D. Bootsma, J.H. Hoeijmakers (1997) Defective transcription-coupled repair in Cockayne syndrome B mice is associated with skin cancer predisposition. Cell *89:* 425–435.

Van Zant, G., B.P. Holland, P.W. Eldridge, J.J. Chen (1990) Genotype-restricted growth and aging patterns in hematopoietic stem cell populations of allophenic mice. J Exp Med *171:* 1547–1565.

Van Zant, G., B.P. Thompson, J.J. Chen (1991) Differentiation of chimeric bone marrow in vivo reveals genotype-restricted contributions to hematopoiesis. Exp Hematol *19:* 941–949.

Vickers, M.A., E. McLeod, T.D. Spector, I.J. Wilson (2001) Assessment of mechanism of acquired skewed X inactivation by analysis of twins. Blood *97:* 1274–1281.

Vijg, J. (2000) Somatic mutations and aging: a re-evaluation. Mutat Res *447:* 117–135.

Wakayama, T., A.C. Perry, M. Zuccotti, K.R. Johnson, R. Yanagimachi (1998) Full-term development of mice from enucleated oocytes injected with cumulus cell nuclei. Nature *394:* 369–374.

Weeda, G., I. Donker, J. de Wit, H. Morreau, R. Janssens, C.J. Vissers, A. Nigg, H. van Steeg, D. Bootsma, J.H. Hoeijmakers (1997) Disruption of mouse ERCC1 results in a novel repair syndrome with growth failure, nuclear abnormalities and senescence. Curr Biol *7:* 427–439.

Wilmut, I., A.E. Schnieke, J. McWhir, A.J. Kind, K.H. Campbell (1997) Viable offspring derived from fetal and adult mammalian cells [see comments]. Nature *385:* 810–813.

Cells
Tissues
Organs

Cells Tissues Organs 2002;171:38–43

Plasticity of Pax-5$^{-/-}$ Pre-B I Cells

Ludovica Bruno[a,b] Christoph Schaniel[a,c] Antonius Rolink[a,d]

[a]Basel Institute for Immunology, Basel, Switzerland; [b]Gene Function and Regulation, Institute of Cancer Research, London, UK; [c]Department of Molecular Biology, Lewis Thomas Laboratory, Princeton University, Princeton, N.J., USA; [d]Department of Immunology, University of Basel, Basel, Switzerland

Key Words
Hematopoiesis · Stem cell · Self-renewal · Differentiation

Abstract
In Pax-5 deficient mice, B cell development is blocked at the pre-B I cell stage. In contrast to their wild type counterparts, Pax-5$^{-/-}$ pre-B I cells posses an extraordinary developmental plasticity showing hematopoietic stem cell features such as pluripotency and self renewing capacity. Here we review the in vitro and in vivo evidence of this intriguing phenomenon.

Introduction

A rare bone marrow cell is responsible for the sustained production of the different cell types present in blood [Weissman, 2000]. T lymphocytes, B lymphocytes, natural killer (NK) cells (cells of the lymphoid lineage) and erythrocytes, megakaryocytes, platelets, granulocytes, monocytes, macrophages, osteoclasts and dendritic cells (cells of the erythroid/myeloid lineages) develop all from a pluripotent hematopoietic stem cell (HSC). This pluripotent HSC is characterized by two capacities, the potential to give rise to all cells of the blood and the capacity to self-renew. Recently, it has been reported that HSCs generate brain, liver and skeletal muscle cells [Theise et al., 2000a; Lagasse et al., 2000; Ferrari et al., 1998; Gussoni et al., 1999] On the other hand, muscle and neural stem cells have been reported to generate blood cells [Jackson et al., 1999; Bjornson et al., 1999]. Thus, HSCs as well as other stem cells appear to have a high degree of developmental plasticity.

The differentiation of HSCs into the various blood lineages is conceived in a hierarchical fashion in which these cells first develop into progenitors and than in precursors following a scheme of inverse relation between decreasing of pluripotency and increasing of commitment to the single differentiation pathways. This model is supported by the identification of a common lymphoid progenitor (CLP) [Kondo et al., 1997] and a common myeloid progenitor (CMP) [Akashi et al., 2000]. A CLP can differentiate into T, B and NK cells, but lost the potential to differentiate into myeloid cells, while a CMP can give rise to various cells belonging to the myeloid lineage, but not to cells of the lymphoid lineage.

Moreover, progenitors and intermediate stages of hematopoietic development have been shown to exhibit some plasticity in their developmental program when they are genetically modified. Thus, Kondo and colleagues [2000] have recently shown that when CLPs are transfected with the IL-2 receptor β-chain gene they can, under appropriate in vitro conditions, differentiate into cells of the myeloid lineage. The bipotential B lymphocyte-macrophage progenitor isolated from bone marrow [Montecino-Rodriguez et al., 2001] is another example of plasticity of a more differentiated hematopoietic precursor.

Perhaps one of the greatest plasticity of this kind yet demonstrated is the in vitro and in vivo differentiation of

KARGER
Fax + 41 61 306 12 34
E-Mail karger@karger.ch
www.karger.com

1422–6405/02/1711–0038$18.50/0

Accessible online at:
www.karger.com/journals/cto

Ludovica Bruno
Section of Gene Function and Regulation
Chester Beatty Labs., Institute of Cancer Research
237 Fulham Road, London SW3 6JB (UK)
Tel. +44 207 970 6114, Fax +44 207 970 6082, E-Mail lbruno@icr.ac.uk

Pax-5-deficient pre-B I cells into various hematopoietic lineages [Nutt et al., 1999a; Rolink et al., 1999b].

In this review, we will discuss the developmental features of Pax-$5^{-/-}$ pre-B I cell clones both in vitro, following different culture conditions and in vivo, upon transplantation into various hosts.

B Lymphopoiesis

B lymphopoiesis is an extremely useful tool to examine the process of lineage specification, as its early stages have been finely characterized by expression of surface markers, the rearrangements status of the immunoglobulin heavy chain (IgH) locus and the requirements for growth factors [Hardy et al., 1991; Rolink et al., 1999a]. The earliest B cell progenitor, called pre-B I, is characterised by D_H to J_H rearrangements and the cell surface expression of B220, CD19 and the tyrosine kinase receptor c-kit [Rolink et al., 1999a]. Pre-B I cells express IL-7 receptor and proliferate in vitro on stromal cells in the presence of IL-7 [Rolink et al., 1991]. Upon removal of IL-7 from the culture system, differentiation of pre-B I cells into $sIgM^+$ immature B cells can be induced [Rolink et al., 1991]. Moreover, transplantation of in vitro grown pre-B I cells into $RAG^{-/-}$, lymphocyte-deficient mice, results in a partial reconstitution of the B cell compartment. Thus, wild type pre-B I cells can both in vitro and in vivo give rise to B cells. However, and in marked contrast to Pax-$5^{-/-}$ pre-B I cells discussed below, wild type pre-B I cells were never found to be able to differentiate into hematopoietic lineages other than B lymphocytes.

The Role of Pax-5 in B Cell Commitment

The mechanisms controlling the commitment of hematopoietic progenitor cells to the different lymphoid lineages are still mostly unknown. Recent findings indicate that the earliest phase of B cell development may proceed in two steps. At the onset of B lymphopoiesis, the transcription factors E2A and EBF coordinately activate the B-cell-specific gene expression program [Sigvardsson et al., 1997; Kee et al., 1998]. Subsequently, Pax-5 appears to repress the promiscuous transcription of lineage-inappropriate genes [Nutt et al., 1999a] and activates B cell lineage specific genes [Nutt et al., 1998; Lauring et al., 1999; Kishi et al., 2000].

The Pax family of transcription factors has been shown to play a crucial role in organogenesis and development, examples are Pax-6 for eye development or Pax-2 and Pax-9 for thymic anlage [Mansouri et al., 1999; Gehring and Ikeo, 1999; Peters et al., 1998]. The Pax-5 gene, encoding the B-cell-specific activator protein (BSAP), is expressed within the hematopoietic system exclusively in the B-lymphoid lineage [Barberis et al., 1990; Li et al., 1996]. Pax-5-deficient mice show a complete block in B cell development at the pre-B I cell stage. Pax-$5^{-/-}$ pre-B I cells, like their wild type counterparts, express various lymphoid and/or B cell-specific genes such as Rag1, Rag2, Igα (mb-1), Igβ (B29), V-pre-B and λ5 as well as sterile transcripts of the IgH locus. Moreover, also Pax-$5^{-/-}$ pre-B I cells posses the long-term capacity to grow in vitro on stromal cells in the presence of IL-7 and can be cloned as individually D_HJ_H-rearranged cell clones [Urbanek et al., 1994; Nutt et al., 1997]. Whereas withdrawal of IL-7 from wild type pre-B I cell cultures results in the development of $sIgM^+$ immature and mature B cells [Rolink et al., 1991], this differentiation is blocked for Pax-$5^{-/-}$ pre-B I cells.

In marked contrast to wild type, Pax-$5^{-/-}$ pre-B I cells also express genes of other lineage-specific programs. These include genes encoding for the M-CSF receptor, myeloperoxidase, perforin, pre-Tα, Epo-R and GATA-1. Interestingly, several of these genes are efficiently repressed by restoration of Pax-5 activity in these cells [Nutt et al., 1999a].

Hematopoietic Pluripotency of Pax$5^{-/-}$ Pre-B I Cells

Pre-B I cells directly isolated from bone marrow of Pax-5-deficient mice, as well as single-cell clones grown in vitro, show extraordinary developmental plasticity, in both in vitro and in vivo systems.

In vitro differentiation experiments demonstrated that Pax-$5^{-/-}$ pre-B I cells have characteristics of an uncommitted hematopoietic progenitor cell with a broad lympho-myeloid development potential. In particular, substitution of IL-7 with lineage-appropriate cytokines allowed Pax-$5^{-/-}$ pre-B I cells to differentiate to functional macrophages (when cultured with M-CSF), osteoclasts (in the presence of TRANCE), dendritic cells (DCs) (with GM-CSF), granulocytes (with G-CSF) and natural killer (NK) cells (with IL-2) [Nutt et al., 1999b] (fig. 1).

Initial in vivo transplantation experiments showed that in vitro grown Pax-$5^{-/-}$ pre-B I cells can fully reconstitute the thymus and the peripheral T cell pool in RAG-$2^{-/-}$ mice and in part the osteoclast compartment in

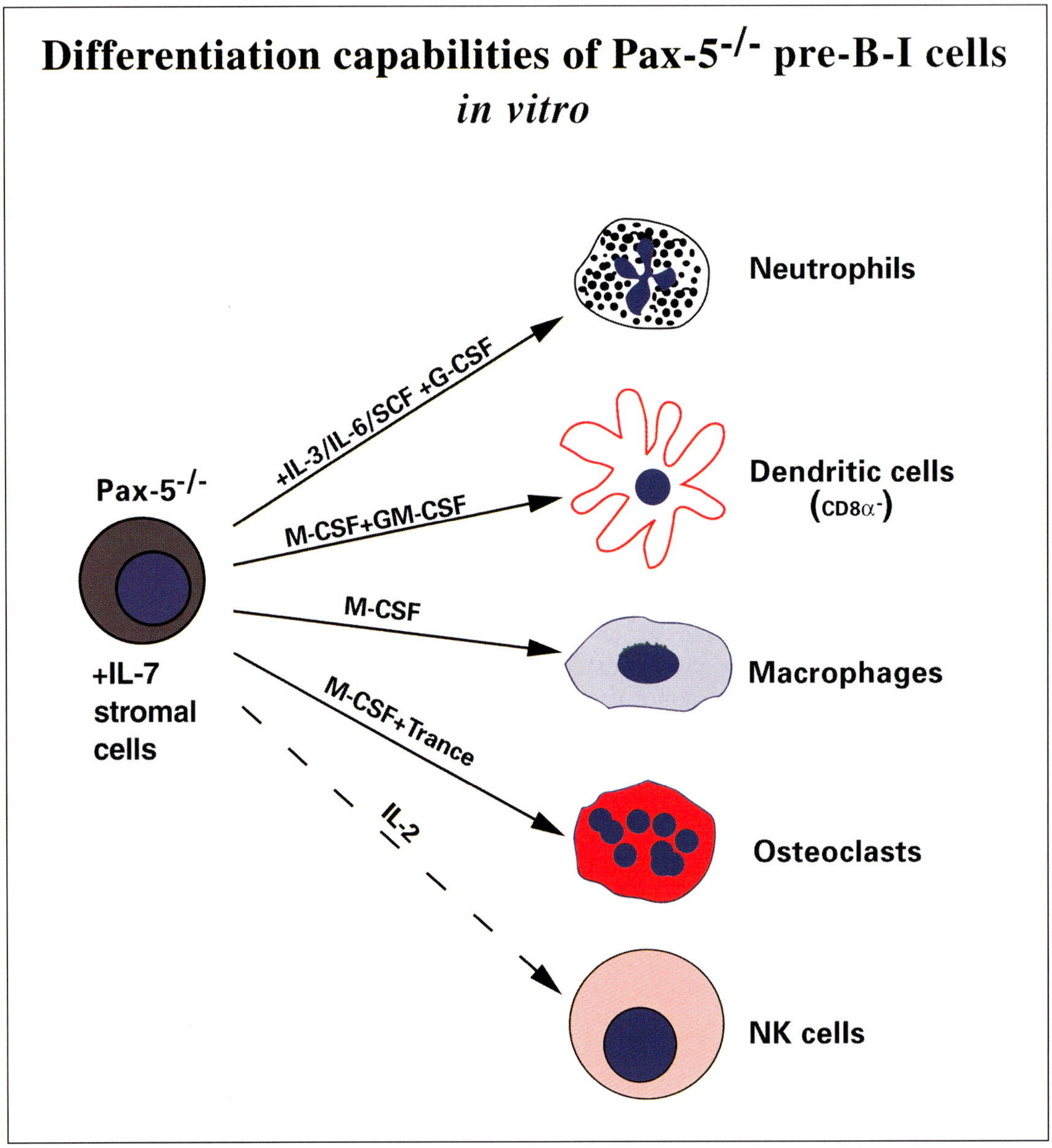

Fig. 1. Schematic representation of Pax-$5^{-/-}$ pre-B I cell differentiation capabilities in vitro.

c-$fos^{-/-}$ mice [Nutt et al., 1999a; Rolink et al., 1999b]. Since the other hematopoietic cell lineages, into which Pax-$5^{-/-}$ pre-B I cells could develop in vitro, were not detected in these in vivo transplantation experiments, it was argued that Pax-$5^{-/-}$ pre-B I cells could only populate those hematopoietic compartments for which the host carries a genetic deficiency resulting in the absence or reduction in numbers of that given hematopoietic cell lineage. However, very recently we demonstrated that Pax-$5^{-/-}$ pre-B I cell clones can also repopulate hematopoietic cell lineages for which the host's endogenous progenitors and mature cell compartments are not deficient. Thus, we showed that transplantation of Pax-$5^{-/-}$ pre-B I cells clones into RAG-$2^{-/-}$ mice results in the development of NK cells, dendritic cells (DCs), macrophages and granulocytes, all lineages in which RAG-$2^{-/-}$ host has no deficiencies [Bruno et al., 2002]. It should however be noted that the developmental kinetics of these cell lineages are much slower than that of T cells and, therefore, have been missed in the previous studies.

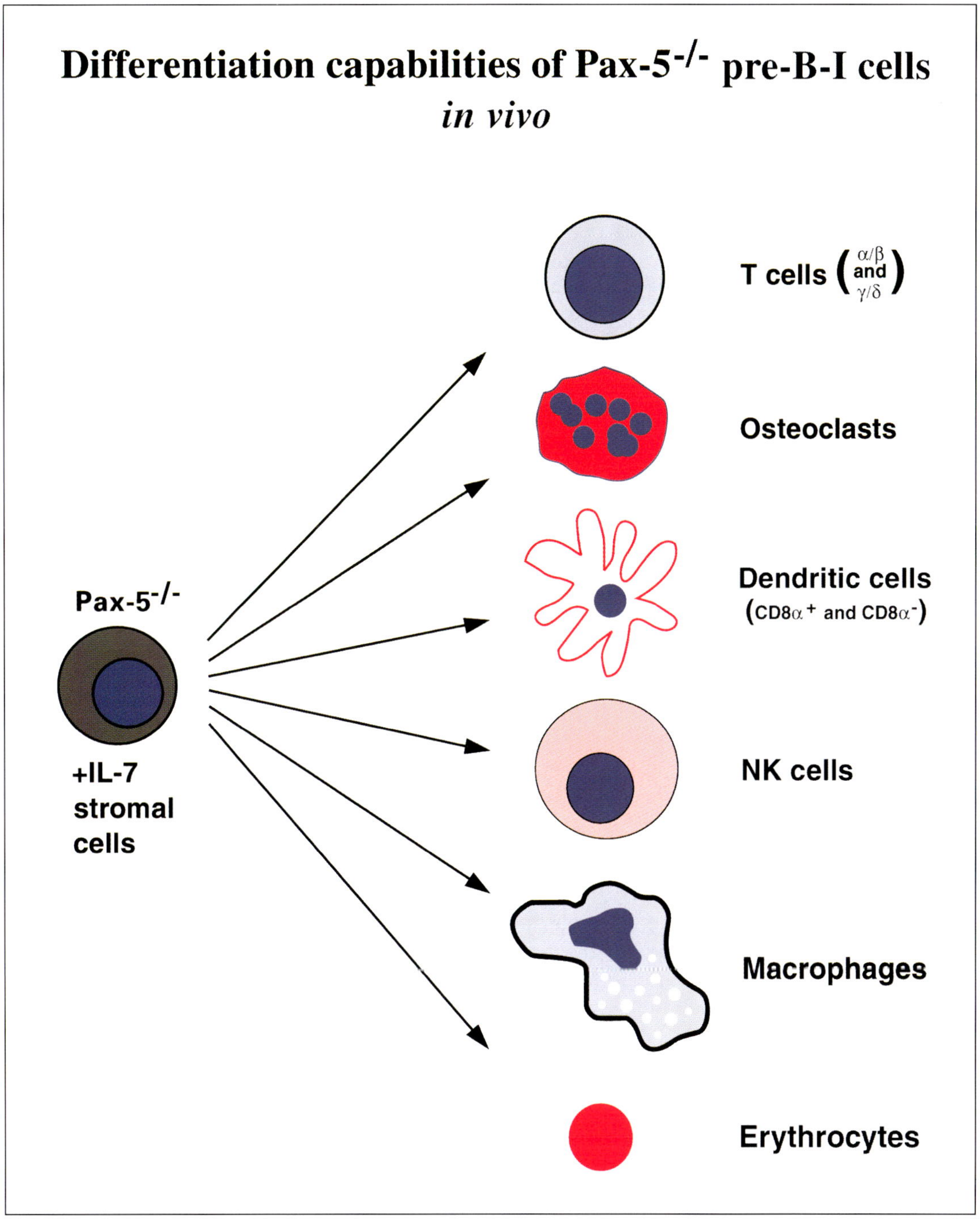

Fig. 2. Schematic representation of Pax-$5^{-/-}$ pre-B I cell developmental plasticity in vivo.

In fact, the in vivo developmental potential of Pax-$5^{-/-}$ pre-B I cells even exceeds the previously observed in vitro. Thus, in vivo Pax-$5^{-/-}$ pre-B I cells can give rise to both CD8α^- and CD8α^+ DCs while in vitro only the development of CD8α– was found. Moreover, in about 10% of transplanted RAG-$2^{-/-}$ mice low numbers (maximally 3%) of erythrocytes of Pax-$5^{-/-}$ origin could be detected.

It should also be stressed here that practically all Pax-$5^{-/-}$ pre-B I cell clones tested show in vitro and in vivo this high degree of plasticity. However, starting with a homogeneous, cloned cell population, the differentiation efficiency in vitro into the various hematopoietic cell lineages is poor. The in vivo differentiation efficiency has thus far not been addressed.

Thus the plasticity of the Pax-5$^{-/-}$ pre-B I cells is cell intrinsic and moreover is in vivo much greater than previously anticipated (fig. 2).

Self Renewal of Pax5$^{-/-}$ Pre-B I Cells

Pluripotency and long-term self-renewal capacity are the characteristics of pluripotent hematopoietic stem cells (pHSC). Above, we summarized the hematopoietic pluripotency of Pax-5$^{-/-}$.

In a first set of experiments, it was shown that Pax-5$^{-/-}$ pre-B I cells also possessed the second feature of pHSC, self renewal. Thus, upon transplantation into RAG-2$^{-/-}$ hosts, Pax-5$^{-/-}$ pre-B I cells reconstituted the pre-B cell compartment in bone marrow [Rolink et al., 1999b]. These pre-B I cells could be isolated and grown in vitro without loosing their in vitro and in vivo multiple hematopoietic lineage developmental potential.

More recently, the extent of this self-renewing capacity was analysed in a serial transplantation approach. Between each transplantation, single Pax-5$^{-/-}$ pre-B I cells were sorted from the bone marrow of RAG-2$^{-/-}$ recipients and grown in vitro. The number of cell divisions each clone underwent in vitro was estimated by cell counting. Thus far, Pax-5$^{-/-}$ pre-B I clones have been five times serially transplanted without loosing their pluripotency and self renewal capacity. During the in vitro expansion period these cells had divided at least 200 times (T.R., unpubl. observation).

Thus, Pax-5$^{-/-}$ pre-B I cells posses a long-term self renewal capacity.

Concluding Remarks

The unexpected extraordinary plasticity of Pax-5$^{-/-}$ pre-B I cells will entitle a series of important investigations. First, the introduction of genes by retroviral transfection will allow to study their effects on different developmental programs. Second, Pax-5$^{-/-}$ pre-B I cells carrying a second mutation not permitting to develop into a specific hematopoietic lineage in which wild type Pax-5$^{-/-}$ pre-B I cells normally would develop, will enable us to identify genes that can rescue this developmental potential. In this context, it is noteworthy that the incapacity of Pax-5$^{-/-}$/SLP-76 double-deficient pre-B I cells to give rise to T cells in vivo can be rescued by retroviral introduction of wild type SLP-76 (A.R., unpubl. observation).

Third, the capacity of Pax-5$^{-/-}$ pre-B I cells to develop, under defined conditions, into various hematopoietic lineages will help to identify genes playing important roles in these different developmental programs. Recently, we have established a culture system that allows very efficiently the differentiation of Pax-5$^{-/-}$ pre-B I cells into DCs. Various gene expression profiling techniques are currently being used on these cells in order to identify genes that might play a role in DC development (L.B., unpubl. observation).

An important and not yet answered question is whether a cell with a similar in vivo plasticity as the Pax-5$^{-/-}$ pre-B I cells is present in normal bone marrow. If this is the case, it might offer an unique system to correct haematological deficiencies in vivo. Therefore, the identification of a population of cells with similar features especially in human, and the establishment of a culture system that keeps their developmental plasticity, is a major future challenge.

References

Akashi, K., D. Traver, T. Miyamoto, I.L. Weissman (2000) A clonogenic common myeloid progenitor that gives rise to all myeloid lineages. Nature *404:* 193–197.

Barberis, K., L. Widenhorn, L. Vitelli, M. Busslinger (1990) A novel B-cell lineage-specific transcription factor present at early but not late stages of differentiation. Genes Dev *5:* 849–859.

Bjornson, C.R., R.L. Rietzle, B.A. Reynolds, M.C. Magli, A.L. Vescovi (1999) Turning brain into blood: hematopoietic fate adopted by adult neural stem cells in vivo. Science *283:* 534–537.

Bruno, L., C. Schaniel, F. Melchers, A. Rolink (2002) Multiple hematopoietic cell lineages develop in vivo from transplanted Pax5-deficient pre-B I cell clones. Blood, in press.

Ferrari, G., G. Cusella-de Angelis, M. Coletta, E. Paolucci, A. Stornaiuolo, G. Cossu, F. Mavilio (1998) Muscle regeneration by bone marrow-derived myogenic progenitors. Science *279:* 1528–1530.

Gehring, W.J., K. Ikeo (1999) Pax6: mastering eye morphogenesis and eye evolution. Trends Genet *9:* 371–377.

Gussoni, E., Y. Soneoka, C.D. Strickland, E.A. Buzney, M.K. Khan, A.F. Flint, L.M. Kunkel, R.C. Mulligan (1999) Dystrophin expression in the mdx mouse restored by stem cell transplantation. Nature *243:* 390–394.

Hardy, R.R., C.E. Carmack, S.A. Shinton, J.D. Kemp, K. Hayakawa (1991) Resolution and characterisation of pro-B and pre-pro-B cell stages in normal mouse bone marrow. J Exp Med *173:* 1213–1225.

Jackson, K.A., T. Mi, M.A. Goodell (1999) Hematopoietic potential of stem cells isolated from murine skeletal muscle. Proc Natl Acad Sci USA *96:* 14482–14486.

Kee, B.L., C. Murre (1998) Induction of early B cell factor (EBF) and multiple B lineage genes by the basic helix-loop-helix transcription factor E12. J Exp Med *188:* 699–713.

Kishi, H., X.C. Wei, Z.X. Jin, Y. Fujishiro, T. Nagata, T. Matsuda, A. Muraguchi (2000) Lineage specific regulation of the murine RAG-2 promoter: GATA-3 in T cells and Pax-5 in B cells. Blood *95:* 3845–3852.

Kondo, M., I.L. Weissman, K. Akashi (1997) Identification of clonogenic common lymphoid progenitors in mouse bone marrow. Cell *91:* 661–672.

Kondo, M., D.C. Sherer, T. Miyamoto, A.G. King, K. Akashi, K. Sugamura, I.L. Weissman (2000) Cell fate conversion of lymphoid-committed progenitors by instructive actions of cytokines. Nature *407:* 383–386.

Lagasse, E., H. Connors, M. Al-Dhalimy, M. Reitsma, M. Dohse, L. Osborne, X. Wang, M. Finegold, I.L. Weissman, M. Grompe (2000) Purified hematopoietic stem cells can differentiate into hepatocytes in vivo. Nature Med *11:* 1229–1234.

Lauring, J., M.S. Schlissel (1999) Distinct factors regulate the murine RAG-2 promoter in B- and T-cell lines. Mol Cell Biol *19:* 2601–2612.

Li, Y.-S., R. Wasserman, K. Hayakawa, R.R. Hardy (1996) Identification of the earliest B lineage stage in mouse bone marrow. Immunity *5:* 527–535.

Mansouri, A., G. Goudreau, P. Gruss (1999) Pax genes and their role in organogenesis. Cancer Res *59:* 1707–1710.

Montecino-Rodriguez, E., H. Leathers, K. Dorshkind (2001) Bipotential B-macrophage progenitors are present in adult bone marrow. Nature Immunol *2:* 84–88.

Nutt, S., P. Urbanek, A. Rolink, M. Busslinger (1997) Essential functions of Pax5 (BSAP) in pro-B cell development: differences between fetal and adult B lymphopoiesis and reduced V-to-DJ recombination at the IgH locus. Genes Dev *11:* 476–491.

Nutt, S.P., A.M. Morrison, P. Dorfler, A. Rolink, M. Busslinger (1998) Identification of BSAP (Pax-5) target genes in early B-cell development by loss- and gain-of-function experiments. EMBO J *17:* 2319–2333.

Nutt, S., B. Heavey, A. Rolink, M. Busslinger (1999a) Commitment to the B lymphoid lineage depends on the transcription factor Pax5. Nature *401:* 556–562.

Nutt, S., A. Rolink, M. Busslinger (1999b) The molecular basis of B-cell lineage commitment. Cold Spring Harb Symp Quant Biol *64:* 51–59.

Peters, H., A. Neubuser, K. Kratochwil, R. Balling (1998) Pax9-deficient mice lack pharyngeal pouch derivatives and teeth and exhibit craniofacial and limb abnormalities. Genes Dev *17:* 2735–2747.

Rolink, A., A. Kudo, H. Karasuyama, Y. Kikuchi, F. Melchers (1991) Long term proliferating early pre B cell lines and clones with the potential to develop to surface positive Ig-positive, mitogen reactive B cells in vitro and in vivo. EMBO J *10:* 327–336.

Rolink, A., F. Melchers, J. Andersson (1996) The SCID but not the RAG-2 gene product is required for su-Se heavy chain class switching. Immunity *5:* 319–330.

Rolink, A., E. ten Boeckel, T. Yamagami, R. Ceredig, J. Andersson, F. Melchers (1999a) B cell development in the mouse from early progenitors to mature B cells. Immunol Letters *68:* 89–93.

Rolink, A., S.L. Nutt, F. Melchers, M. Busslinger (1999b) Long-term in vivo reconstitution of T-cell development by Pax5-deficient B-cell progenitors. Nature *401:* 603–606.

Sigvardsson, M., M. O'Riordan, R. Grosschedl (1997) EBF and E47 collaborate to induce expression of the endogenous immunoglobulin surrogate light chain genes. Immunity *7:* 25–36.

Theise, N.D., M. Nimmakayalu, R. Gardner, P.B. Ilei, G. Morgan, L. Teperman, O. Henegariu, D.S. Krause (2000a) Liver from bone marrow in humans. Hepatology *32:* 11–16.

Theise, N.D., S. Badve, R. Saxena, O. Henegariu, S. Sell, J.M. Crawford, D.S. Krause (2000b) Derivation of hepatocytes from bone marrow cells in mice after radiation-induced myeloablation. Hepatology *31:* 235–240.

Urbanek, P., Z.-Q. Wang, I. Fetka, E.F. Wagner, M. Busslinger (1994) Complete block of early B cell differentiation and altered patterning of the posterior midbrain in mice lacking Pax5/BSAP. Cell *79:* 901–912.

Weissman, I.L. (2000) Stem cells: units of development, units of regeneration and units in evolution. Cell *100:* 157–168.

Cells
Tissues
Organs

Cells Tissues Organs 2002;171:44–63

Avian Models to Study the Transcriptional Control of Hematopoietic Lineage Commitment and to Identify Lineage-Specific Genes

Julie S. Nielsen Regis Doyonnas Kelly M. McNagny

The Biomedical Research Centre, University of British Columbia, Vancouver, Canada

Key Words
Transcription factor · Avian retrovirus · Podocalyxin · CD34 · Adhesion

Abstract
E26 is an avian acute leukemia virus with a profound ability to transform multipotent hematopoietic progenitor cells both in vivo and in vitro. Progenitor cells transformed by this virus can be expanded in vitro as undifferentiated clones for up to two months and can also be induced to differentiate into cells of the erythroid, eosinophilic, thrombocytic, and myelomonocytic lineages with reproducible kinetics. Aside from the proliferative stimulus provided by the E26 oncoprotein, these cells are remarkably similar to normal hematopoietic progenitors. They therefore provide an ideal assay system for determining the influence of ectopically expressed transcription factors on both maturation and commitment to several hematopoietic lineages. Results from experiments using this system suggest that subtle shifts in the balance of lineage-restricted transcription factors can result in profound changes in phenotype and challenge the notion that lineage commitment is a uni-directional process. Analysis of the regulatory elements governing the expression of these genes has provided novel mechanistic insights into the transcriptional control of hematopoiesis. In addition to their utility in deciphering the control of lineage commitment, the ability to grow large numbers of undifferentiated and more mature hematopoietic cells has facilitated the discovery of a number of novel, lineage-restricted genes. Analysis of the proteins encoded by these genes is helping to clarify the role of a number of membrane proteins in the interaction between hematopoietic cells and their microenvironments.

Abbreviations used in this paper

MEP	Myb-Ets transformed progenitors
PKC	protein kinase C
IRES	internal ribosome entry site
FOG	friend of GATA
C/EBP	CCAAT/enhancer-binding protein
CKII	casein kinase II
TK	tyrosine kinase

KARGER

Fax + 41 61 306 12 34
E-Mail karger@karger.ch
www.karger.com

1422–6405/02/1711–0044$18.50/0

Accessible online at:
www.karger.com/journals/cto

Kelly M. McNagny
Biomedical Research Centre, University of British Columbia
Vancouver, British Columbia, V6T 1Z3 (Canada)
Tel. +1 604 822 7810, Fax +1 604 822 7815, E-Mail Kelly@brc.ubc.ca

Introduction

How does a hematopoietic stem cell decide what to become? Is this decision stochastic or directed? These questions have been a topic of research and discussion for many years. Whether the decision is random or determined by external factors, key transcription factors are certainly involved in executing this decision and regulating expression of the lineage-specific genes that determine a cell's phenotype. What are the key transcriptional regulators? Is there a single master regulator of each lineage, or does commitment require the interaction of combinations of transcription factors, working together to activate and repress lineage-specific genes?

There are clearly many questions regarding hematopoiesis, but how can they best be addressed? One approach involves expression of individual candidate transcription factors in normal hematopoietic progenitor cells. Unfortunately, these progenitor cells are very rare, and even if a large enough quantity could be isolated, they are difficult to maintain and expand in vitro due to their intrinsic tendency to differentiate. An alternative approach has been to express candidate regulators in hematopoietic progenitor cell lines. However, the mechanism behind the immortalization of cell lines is often unclear, their gene expression patterns can be quite aberrant, and many have lost the capacity to respond to normal differentiation signals.

The E26 system is a useful alternative for studying hematopoiesis. E26 is an avian oncogenic retrovirus that induces an acute hematopoietic progenitor cell leukemia in birds. In vitro the virus transforms multipotent hematopoietic progenitors (called MEPs for Myb-Ets-transformed multipotent progenitors) and myeloblasts (fig. 1) [Graf et al., 1992]. Although most acute leukemia viruses allow the analysis of differentiation along just a single lineage, the E26 system permits the study of differentiation along myeloid, erythroid, eosinophilic, and thrombocytic lineages [Graf et al., 1992; McNagny and Graf, 1997]. In this report we will review the components of the E26 system and its utility in identifying transcription factors that regulate hematopoietic lineage commitment and maturation. In addition, we will describe how this system has been used to identify novel Myb- and Ets-regulated transcripts and to clone two novel lineage-specific cell surface proteins. Although the latter antigens were initially used exclusively as phenotypic markers, the subsequent cloning of these membrane proteins, the characterization of their promoters for lineage-specific transcriptional elements, and the mutagenesis of their coding sequences are beginning to provide valuable insights into the regulation of normal hematopoietic development.

The E26 System

The transforming entity of the E26 virus consists of a unique 135 kDa protein encoding a portion of the retroviral Gag sequences fused in-frame to mutant forms of 2 cellular transcription factors: c-Myb and c-Ets-1 [Leprince et al., 1983; Nunn et al., 1983; reviewed in McNagny and Graf, 1997 and Sieweke and Graf, 1998]. In vitro transformation of multipotent progenitors requires activating mutations and truncations in both c-Myb and c-Ets-1, physical linkage of these 2 transcription factors, intact Myb and Ets-1 DNA-binding domains, and the Myb transactivation domain (fig. 1A) [Metz and Graf, 1991a, b].

The E26 system is quite simple to use and has been described in detail previously [Graf et al., 1992; McNagny and Graf, 1997]. Briefly, 2-day chick embryos (12–18 somite blastoderm) are infected by culturing with viral stocks or by co-cultivation with a packaging cell line producing the virus (fig. 1B). Colony assays are then performed in methylcellulose-containing medium. These cultures contain small colonies of uninfected erythroid cells and large colonies of infected cells. Colonies are picked after 7–10 days and expanded in liquid culture. These cells proliferate for 3–6 weeks before beginning to senesce. E26 transformation results in 2 types of colonies: myeloblast colonies and multipotent progenitor colonies (MEPs). The myeloblast colonies probably arise due to the spontaneous differentiation of progenitors within the colony assay, and they typically represent about 30% of the total colonies [Graf et al., 1992]. The actual frequency of myeloblast colonies is influenced by duration of co-cultivation with packaging cells: longer co-cultivations increase myeloblast frequency. MEP cells are largely blocked in differentiation, but they can be induced to differentiate at high frequency along the myeloid or eosinophil lineages upon exposure to phorbol esters (such as TPA) or by activation of various other kinase cascades (fig. 1C) [Graf et al., 1992]. The intracellular receptor for TPA is protein kinase C (PKC) which is a component of signal transduction pathways involved in maturation and commitment. Results from the E26 system suggest that a lack of PKC activity maintains the progenitor phenotype while low PKC activity (high TPA) leads to myelomonocytic differentiation and high PKC activity (low TPA) results in predominantly eosinophil differentiation [Rossi et al., 1996b]. Eosinophilic cells in this system tend to terminally differentiate and become postmitotic, while

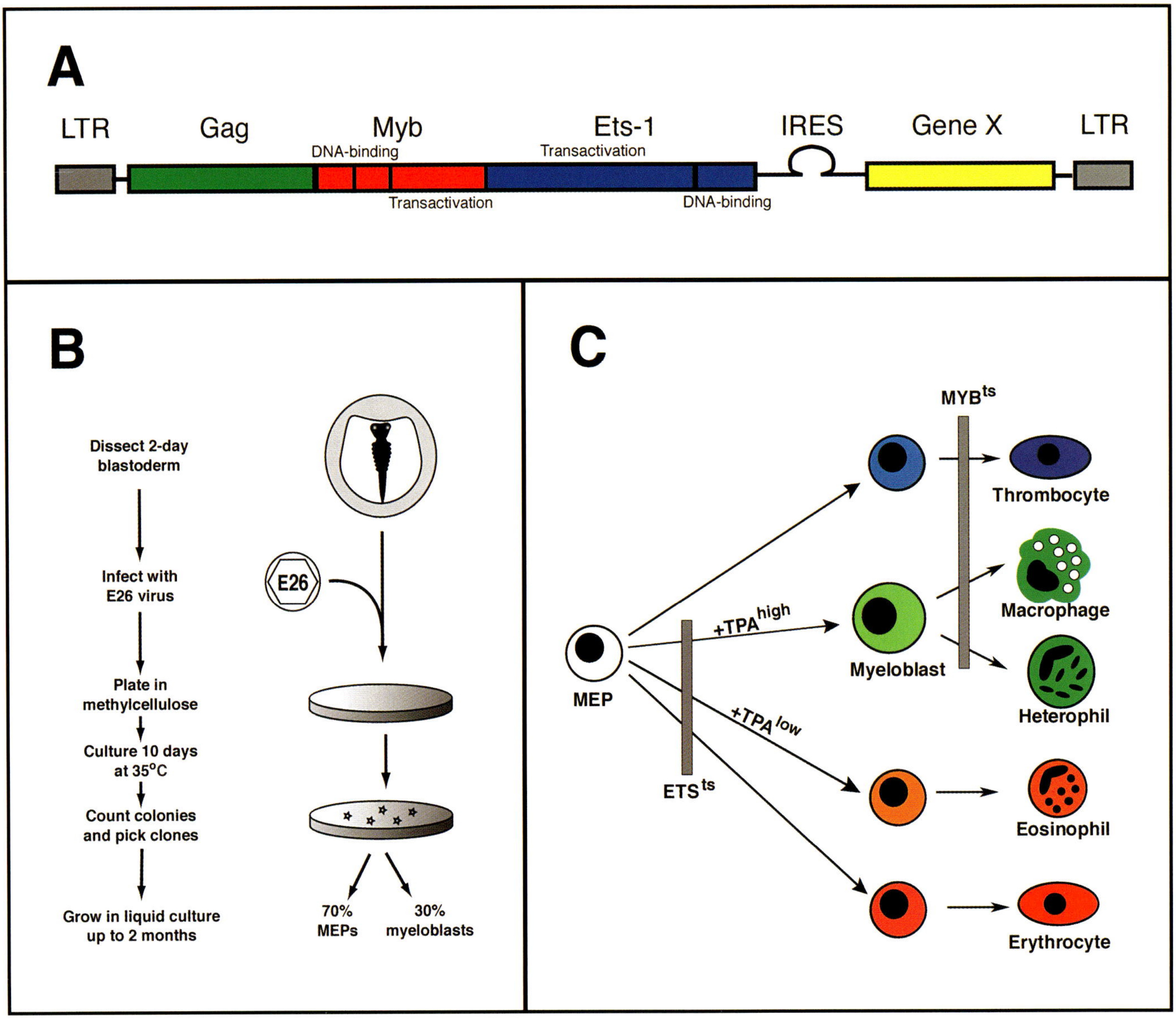

Fig. 1. Schematic overview of the E26 transformation system. **A** Architecture of E26 retroviral vectors. Gray boxes = viral long terminal repeats; green boxes = retroviral gag sequences; red boxes = v-Myb sequences; blue boxes = v-Ets-1 sequences; yellow boxes = coding sequences for any gene of interest; IRES indicates internal ribosomal entry site governing translation of the gene of interest. **B** Schematic representation of blastoderm transformation assay. **C** Differentiation potential of E26-transformed cells. Gray bars indicate blocks in differentiation imposed by the respective Myb and Ets portions of the oncoprotein that can be relieved by temperature shift of temperature-sensitive variants of the virus.

myelomonocytic cells remain as proliferating, undifferentiated granulocyte and macrophage precursors [Beug et al., 1984; Frampton et al., 1996; Graf et al., 1992]. The cell types generated using this system can be easily distinguished based on expression of various cell-surface and intracellular antigens [McNagny and Graf, 1996].

In order to make this system more versatile, temperature-sensitive versions of the E26 virus have been generated which allow temperature-dependent inactivation of the v-Myb or v-Ets DNA-binding domains. One such temperature-sensitive virus, called ts1.1 (Ets^{ts}), has a histidine to aspartic acid mutation in the DNA-binding domain of Ets-1, which completely inhibits its DNA-

Table 1. Factors expressed in E26-transformed cells and their effects on phenotype

Factor	Effect	Reference
v-raf (serine/threonine kinase)	MEPs→myeloblasts/eosinophils*	Graf et al., 1992
v-H-ras (GTPase)	MEPs→myeloblasts/eosinophils*	Graf et al., 1992
v-src (non-receptor-type tyrosine kinase)	MEPs→myeloblasts/eosinophils*	(Graf et al., 1992)
v-sea (truncated receptor-type tyrosine kinase)	MEPs→myeloblasts/eosinophils*	Graf et al., 1992
v-erbA	no effect	Graf et al., 1992
v-erbB	MEPs→myeloblasts/eosinophils	Graf et al., 1992
GATA-1	myeloblasts→eosinophils myeloblasts→MEPs	Kulessa et al., 1995
PKCα	MEPs→myeloblasts/eosinophils	Rossi et al., 1996b
PKCε	MEPs→myeloblasts/eosinophils	Rossi et al., 1996b
bcl-2	ts21 myeloblasts→neutrophil granulocytes (probably by blocking granulocyte apoptosis)	Frampton et al., 1996
PU.1	MEPs→myeloblasts	Nerlov and Graf, 1998
C/EBPα	MEPs→eosinophils	Nerlov et al., 1998
C/EBPβ	MEPs→myeloblasts/eosinophils	Nerlov et al., 1998
FOG-1	eosinophils→MEPs	Querfurth et al., 2000
MafB	MEPs→myeloblasts ts21 myeloblasts→macrophages (rapidly)	Kelly et al., 2000

Highlighted factors are discussed in detail in this review.
* Probably produces eosinophils (unpublished).

binding capacity at the non-permissive temperature [Golay et al., 1988; Kraut et al., 1994]. Cells transformed with this virus can be maintained as multipotent progenitors at the permissive temperature then shifted to a higher temperature to relieve the differentiation block imposed by Ets-1. Temperature shift results in predominantly erythroid differentiation (up to 90%) within 2 days which makes this virus extremely useful for studying differentiation along the erythroid lineage [Golay et al., 1988; Kraut et al., 1994]. A low degree of temperature-dependent eosinophil and myeloblast differentiation is also observed in these cultures. The fact that Ets imposes a differentiation block on erythroid differentiation was further demonstrated using the FLP recombinase system [Rossi et al., 1996a]. FLP recombinase target sites were introduced into the E26 vector to allow inducible removal of the entire Ets-coding sequence. This led to induction of erythroid differentiation by MEPs.

Another E26 virus variant, called ts21 (Myb^{ts}) has a temperature-sensitive mutation (threonine to arginine) in the DNA-binding domain of v-Myb [Beug et al., 1984; Frykberg et al., 1988]. Temperature shift of multipotent progenitors transformed by this virus leads exclusively to terminal thrombocytic differentiation [Frampton et al., 1995]. Alternatively, temperature shift of Myb^{ts}-transformed myeloblasts leads to differentiation into macrophages and to a lesser extent heterophils (the chicken equivalent of neutrophils [Beug et al., 1984; Frampton et al., 1995]). This virus is therefore particularly useful for analyzing myelomonocytic and thrombocytic differentiation.

Addition of an internal ribosome entry site (IRES) to the E26 viral genome adds further versatility to this system in that it allows co-expression of potential regulators of hematopoietic differentiation [Rossi et al., 1996b]. Primary transformants essentially act as cellular test tubes for analyzing the function of these regulators. To date, over a dozen transcription factors, signaling molecules, membrane proteins, and enzymes have been cloned into the E26 vectors for analysis in this system (table 1). These have provided valuable insights into the regulation of normal hematopoietic differentiation. In the first part of this manuscript, we will review the insights provided by 4 transcription factors or transcriptional regulators: GATA-1, FOG-1, PU.1, and C/EBP.

The E26 System: Insights into Transcriptional Control of Hematopoiesis

GATA-1 Induces Eosinophil Trans-Differentiation or Retro-Differentiation to a Multipotent Phenotype

The zinc finger transcription factor, GATA-1 was initially cloned as an 'erythroid-specific' transcription factor and was shown to be a key regulator of globin gene expression [Evans and Felsenfeld, 1989; Tsai et al., 1989]. Subsequently, it was found to have a much broader expression pattern, and it appears to act as a key regulator of various hematopoietic cell fates [Lemarchandel et al., 1993; Martin et al., 1990; Romeo et al., 1990; Zon et al., 1991]. In the E26 system, GATA-1 is expressed by MEPs, thrombocytic and erythroid cells, and eosinophils [Kulessa et al., 1995; Shivdasani and Orkin, 1996] (fig. 2A). Its expression is notably absent, however, in all cells of the myelomonocytic lineage. This led to the study of its potential influence on the myeloid phenotype in overexpression studies. The first GATA-1 experiments involved introduction of GATA-1 into the myeloblast cell line, HD50M [Kulessa et al., 1995] (fig. 2B). This clonal cell line has been extensively characterized and was initially derived by infection of primary hematopoietic progenitors with E26 followed by TPA-induced myeloblast differentiation. Forced expression of GATA-1 in this cell line produced striking results: clones expressing GATA-1 rapidly lost expression of all myelomonocytic-specific markers and began to express markers specific to either eosinophils or multipotent progenitors (MEPs). Eosinophilic clones acquired expression of eosinophil-peroxidase and the membrane protein EOS47 (see below), both exquisitely specific markers of the eosinophil lineage. MEP clones, in contrast, expressed high levels of progenitor-specific molecules, including MEP21 (a CD34-related molecule, see below and table 1) and MEP26. Interestingly, expression levels of the GATA-1 transgene correlated precisely with clone phenotype: all eosinophilic cells expressed low levels of the GATA-1 transgene, while progenitor cells expressed it at 2–3 fold higher levels.

To ensure that GATA-1 was indeed inducing a phenotypic switch rather than permitting the selective outgrowth of rare HD50M cells that had spontaneously switched phenotype, an inducible GATA-1 construct was generated by fusion to the hormone-binding domain of the estrogen receptor [Kulessa et al., 1995]. GATA-1's ability to bind DNA was thereby rendered estradiol-dependent which allowed accurate analysis of the kinetics of myeloblast differentiation in response to hormone. Upon exposure to estradiol, myeloblast markers were down-regulated and eosinophil markers were up-regulated within 1–2 days (transcripts for some markers were detected as early as 4 h after induction). In contrast, expression of MEP markers required several weeks, which suggests this transition may involve activation of additional nuclear factors [Graf, 2000]. One likely candidate as an additional regulator is the 'Friend of GATA-1' protein, FOG-1 (see below).

The myeloblast-to-MEP transition induced by GATA-1 is a striking result and presented a clear challenge to the dogma that differentiation is a 'one-way street.' To further confirm that the MEP cells induced by ectopic GATA-1 expression had regained multipotentiality, temperature-sensitive mutants were used. As mentioned above, cells transformed by the Ets^{ts} E26 virus (with a temperature-sensitive mutation in the Ets-1 DNA-binding domain) are able to differentiate into erythroid, myeloid, and eosinophilic cells upon temperature shift. Therefore, GATA-1 was introduced into the Ets^{ts}-transformed myelomonocytic cell line, HD13 [Kulessa et al., 1995]. As with HD50M cells, GATA-1 expression in HD13 cells led to a phenotypic shift towards progenitors. Then, after shift to 42°C (which relieves the differentiation block imposed by Ets-1), most clones showed an increase in hemoglobin-positive cells. Detailed analysis of 1 of these clones showed a 10-fold increase in hemoglobin expression, up-regulation of the late erythroblast/erythrocyte marker JS4, and complete down-regulation of the progenitor antigens MEP21 and MEP26.

In summary, the data suggest that low GATA-1 expression levels induce a 'trans-differentiation' of myeloblasts into eosinophils while at high levels GATA-1 can induce a 'retro-differentiation' into multilineage progenitors. It is of interest that expression of threshold levels of GATA-2 in mammalian cells has been shown to regulate the difference between self-renewal and differentiation of progenitors [Heyworth et al., 1999]. Although experiments with GATA-1 were the first major challenge to the dogma that differentiation is unidirectional, similar challenges are now being observed with normal cells in a variety of mammalian systems (see review of Pax5 function in preB cells by Bruno et al. in this issue). Expression levels of genes such as Pax5 and GATA-1 thereby regulate normal versus retro-differentiation.

GATA-1 Binding Cofactors: FOG-1

How is the GATA-1-induced 'retro-differentiation' regulated? What cofactors are required? Recently these questions were addressed using GATA-1 mutants. The E26-transformed myeloblast cell line, HD50M, was trans-

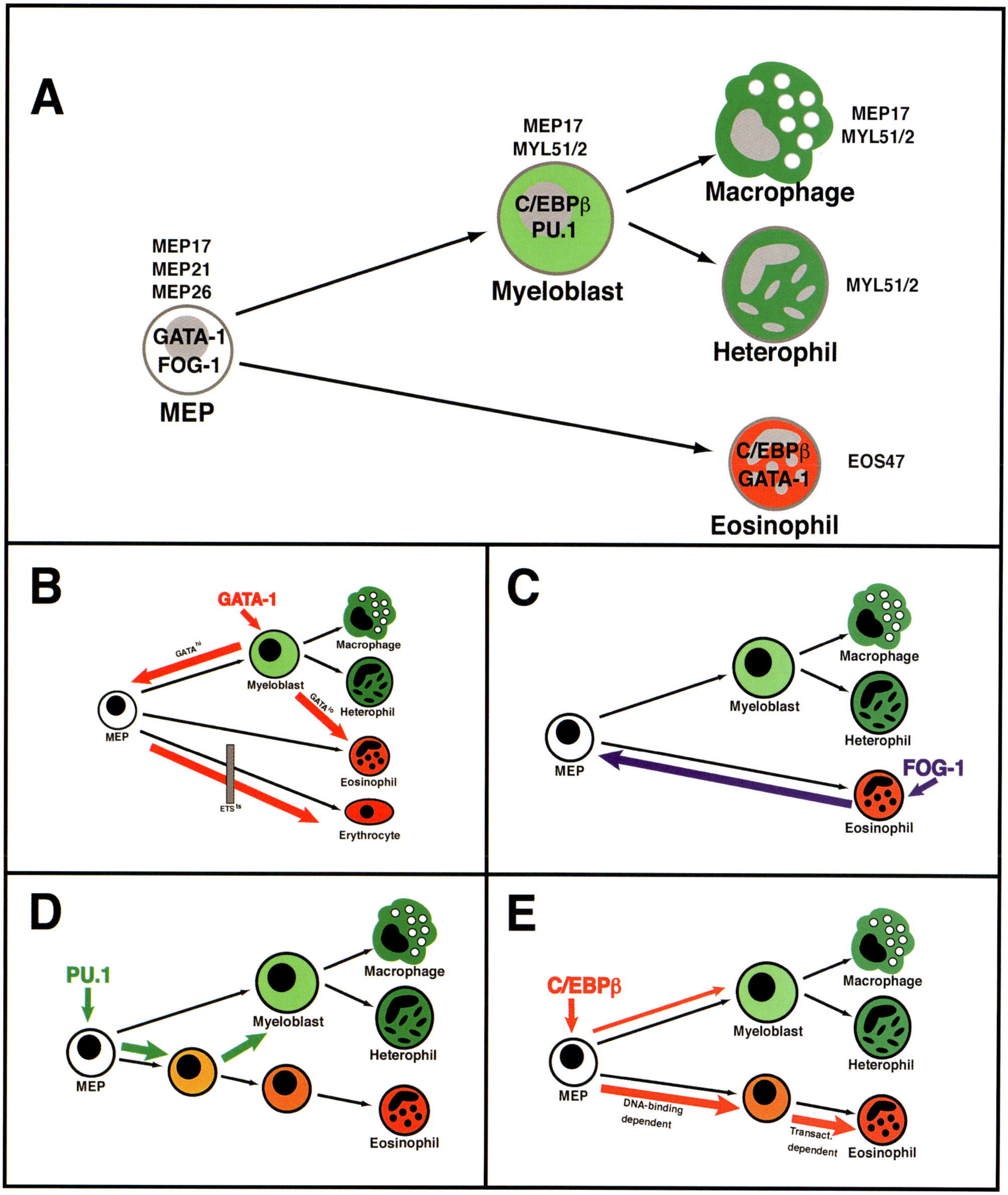

Fig. 2. Schematic showing effects of ectopic transcription factor expression in E26 transformants. **A** The normal expression pattern of transcriptional regulators and antigen markers in E26 transformed cells. **B** Red arrows indicate phenotypic shifts induced by GATA-1 expression in myeloblasts. **C** Blue arrows indicate phenotypic shift induced by FOG-1 expression in eosinophils. **D** Green arrows indicate phenotypic shift induced by PU.1 expression in MEP cells. **E** Orange arrows indicate phenotypic shift induced by C/EBP-type factor expression in MEP cells. Thick lines indicate effects with strong C/EBP-type transactivators; thin lines indicate weaker transactivators.

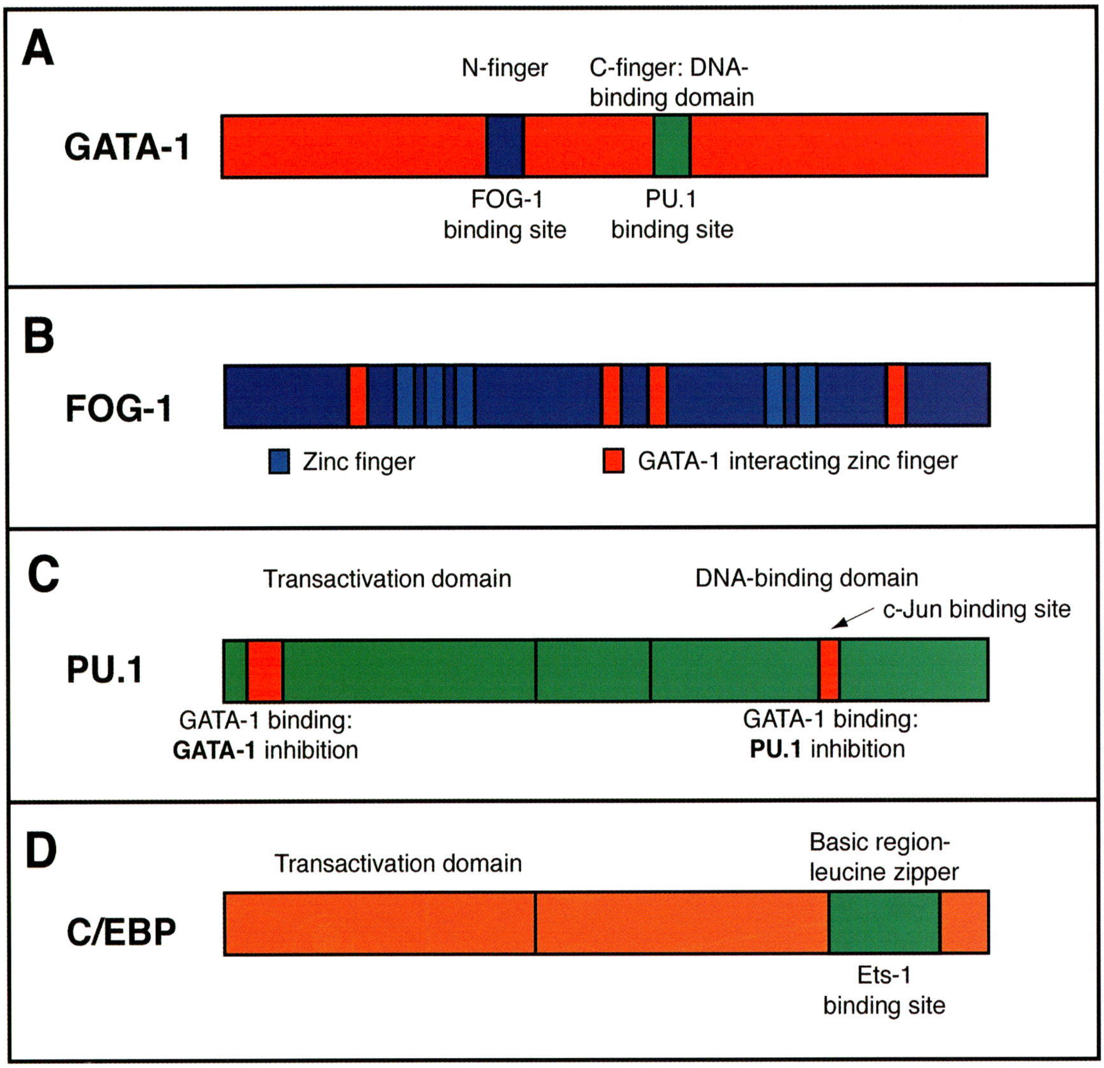

Fig. 3. Schematic of transcription factor domains and binding sites.

fected with wild-type GATA-1 or derivatives in which either the C-terminal zinc finger (DNA-binding domain) or the N-terminal zinc finger were mutated (fig. 3A) [Querfurth et al., 2000]. Interestingly, HD50M cells transfected with the C-terminal DNA-binding mutant failed to differentiate, whereas cells transfected with the GATA-1 N-terminal mutants differentiated into eosinophils. Neither mutant gave rise to progenitor-type cells. This suggests that while C-terminal DNA-binding alone can induce trans-differentiation into eosinophils, the DNA-binding domain and the N-terminal zinc finger are both required for retro-differentiation of myeloblasts into MEPs.

Although the N-terminal domain is not required for specific DNA binding, it does bind to the 'Friend of GATA-1' protein, FOG-1 (fig. 3B). FOG-1 was first identified in a genetic screen for GATA-1 interacting proteins in yeast [Tsang et al., 1997]. It contains 9 zinc fingers, at least 4 of which can bind the N-terminal zinc finger of GATA-1 [Crispino et al., 1999; Fox et al., 1998; Fox et al., 1999; Tsang et al., 1997]. MEPs and erythroid cells were shown to express high levels of FOG-1 mRNA while cells of the myeloid and eosinophil lineages express very low levels [Querfurth et al., 2000]. Interestingly, although GATA-1 was shown to activate the eosinophil-specific EOS47 promoter, co-expression of FOG-1 was shown to inhibit this GATA-1-mediated activation. In addition, transactivation of the eosinophil granule major basic protein by GATA-1 in mammalian cell lines is inhibited by FOG-1 [Yamaguchi et al., 1998, 1999]. However, FOG-1 and GATA-1 act synergistically in mammalian erythroid differentiation [Tsang et al., 1997, 1998]. Thus, the data suggest that FOG-1 is a likely candidate for regulating the eosinophil vs. MEP outcome when GATA-1 is introduced into myeloblasts (fig. 2B). This was tested directly in experiments where FOG-1 was introduced into the eosinophil cell line, 1A1. This cell line was originally derived by GATA-1 expression in HD50M myeloblasts; it expresses high levels of the EOS47 antigen and low levels of eosinophil-peroxidase [Kulessa et al., 1995]. FOG-1 induced the loss of expression of EOS47 as well as eosinophil-peroxidase and the myeloid/eosinophil-specific protein, Mim-1 [Querfurth et al., 2000]. Instead, these cells acquired an MEP morphology and expressed the progenitor markers MEP21 and MEP26 (fig. 2C). In control experiments, the expression of FOG-1 in myeloblasts (which lack GATA-1 expression) did not lead to retro-differentiation toward MEPs, suggesting that the GATA-1/FOG-1 interaction is required for this process to occur. In this in vitro, retro-differentiation model the data argue that GATA-1 and FOG-1 are important for generating the multipotent phenotype. Extrapolating to normal hematopoietic differentiation, the GATA-1/FOG-1 interaction may be required for maintaining the multipotent phenotype, and down-regulation of FOG-1 may represent a key step in eosinophil differentiation.

PU.1 Is an Inducer of Myeloid Differentiation

PU.1 is a transcription factor of the Ets family with a winged helix-turn-helix DNA-binding domain (fig. 3C); it is expressed specifically in myeloid and B-lymphoid cells of the hematopoietic system [Klemsz et al., 1990]. In PU.1 knockout mice these lineages fail to develop or are delayed and highly aberrant [McKercher et al., 1996; Scott et al., 1994]. In order to study the role of PU.1 in hematopoiesis, an E26 construct containing PU.1 expressed from an IRES element was used to infect 2-day chicken blastoderms [Nerlov and Graf, 1998]. In these experiments, wild-type virus infection resulted in 50% MEP clones, 40% myeloblast clones, and 8% eosinophils. In contrast, PU.1 expressing constructs induced a dramatic increase in the number of myeloblast clones as they accounted for greater than 75% of clones, while less than 10% were MEP clones. This suggested that PU.1 is involved in differentiation along the myelomonocytic pathway (fig. 2D). Again, to ensure that this was indeed due to differentiation rather than selective survival of myeloid cells, a conditional allele of PU.1 was created by fusing the hormone-binding domain of the human estrogen receptor (ER) to the C-terminus of human PU.1. This E26 construct was then used to transform blastoderms, and pure MEP clones were isolated. These were then treated with β-estradiol, and the resulting phenotypes were analyzed after 2, 4, and 6 days. A gradual decrease in MEP21 expression and an up-regulation of the myeloid-specific membrane proteins, 1C3 and MYL51/2 were observed. Thus, the increase in myeloblasts observed in these experiments was unlikely to be a result of selection.

Further analysis revealed that after 1 day of estradiol treatment, most cells had down-regulated MEP markers and approximately 10% expressed myeloid markers instead [Nerlov and Graf, 1998]. Interestingly, another subpopulation (consisting of 50% of the cells) expressed EOS47 but was peroxidase-negative, suggestive of an immature eosinophil phenotype. A 3-day β-estradiol treatment led to the full commitment of most cells to the myeloid lineage (cells lacked MEP and eosinophil markers but expressed various myeloid markers, had a myeloblast morphology, and were dependent on chicken myelomonocytic growth factor for proliferation). These cells also acquired expression of the myeloid transcription factor, C/EBPβ (see below), and lost GATA-1 expression. The $EOS47^+$ cells observed early in the experiment suggest the presence of an immature eosinophil intermediate in this pathway. Additional experiments demonstrated that the differentiation of MEPs into myeloblasts is dependent on the amino-terminal transactivation domain of PU.1. Thus, PU.1 induces differentiation along the myeloid lineage via an immature eosinophil intermediate in a manner dependent on its N-terminal transactivation domain.

GATA-1 and PU.1 Are Mutually Antagonistic

GATA-1 and PU.1 have mutually exclusive expression patterns and functions in the hematopoietic system (fig. 2A). GATA-1 is expressed in MEPs, eosinophils, erythroid and thrombocytic lineages, and mast cells [Kulessa et al., 1995; Shivdasani and Orkin, 1996]; overexpression of GATA-1 in myeloblasts leads to eosinophil trans-differentiation and MEP retro-differentiation [Kulessa et al., 1995]. In contrast, PU.1 is expressed specifically in myeloid and B-lymphoid lineages [Klemsz et al., 1990], and its overexpression in MEPs leads to myeloblast differentiation [Nerlov and Graf, 1998]. These observations suggest an antagonistic role for these 2 transcription factors in hematopoiesis [Nerlov and Graf, 1998]. The question then is, 'In any given cell, what determines whether PU.1 or GATA-1 will be dominant?'

To better understand this apparent antagonism, the expression of PU.1 was analyzed in E26-transformed myeloid cell lines where GATA-1 was conditionally activated (through the use of an ER-fusion protein and addition of estradiol) [Nerlov et al., 2000, and see above]. Although GATA-1 activation led to trans-differentiation toward eosinophils, PU.1 mRNA levels remained constant throughout this process. Thus, in this scenario, GATA-1 did not operate by preventing the expression of PU.1 mRNA or protein. This suggested that GATA-1 was inhibiting the function of PU.1 by direct inhibition of PU.1's ability to function as a transcription factor. This was tested using a PU.1-responsive reporter construct. This construct, along with a PU.1 expression vector and increasing amounts of a GATA-1 expression vector, was transfected into fibroblasts. In these experiments, PU.1 activity decreased as GATA-1 expression increased, despite the fact that the amount of PU.1 protein remained constant. This inhibition involved the direct binding of the GATA-1 C-terminal zinc finger to the Ets (DNA-binding) domain of PU.1 (fig. 3). Normally, the Ets domain of PU.1 activates transcription through interactions with c-Jun [Zhang et al., 1999]. Thus, the data are consistent with a model where protein-protein interactions between GATA-1 and PU.1 block the ability of PU.1 to interact with c-Jun and function as a transcription factor.

If GATA-1 inhibits PU.1 function in myeloblasts and induces a phenotypic shift toward eosinophils or progenitors, how does PU.1 expression induce myeloid differentiation in progenitors (which express high levels of GATA-1)? Overexpression studies suggest that PU.1 can also block GATA-1 function [Rekhtman et al., 1999]. Interestingly, PU.1 inhibits GATA-1 DNA-binding activity rather than preventing it from interacting with its cofactor, FOG-1 (fig. 3) [Zhang et al., 2000]. Thus, the data suggest that GATA-1 and PU.1 titrate each other's activity and that it is the relative abundance of each transcription factor that determines cell fate; whichever factor is expressed at higher levels is free to activate lineage-specific genes [Graf, 2000]. These experiments provide an excellent example of how the E26 system can be used to study the complex combinatorial interactions of transcription factors. Similar observations have been made in mammalian myeloid and B cell differentiation [DeKoter and Singh, 2000]. Subtle shifts in the level of PU.1 in normal cells can determine whether differentiation proceeds along the B or myelomonocytic lineage.

C/EBPβ Induces Eosinophil Commitment Followed by Myeloid Commitment

CCAAT/Enhancer-Binding Proteins, or C/EBPs, are transcriptional activators involved in the differentiation and growth arrest of several types of cells [Freytag and Geddes, 1992; Umek et al., 1991]. The members of this family have highly conserved basic DNA-binding and leucine zipper dimerization domains at the carboxy-terminus and divergent regulatory and transactivation domains at the amino-terminus (fig. 3) [Kowenz-Leutz et al., 1994; Landschulz et al., 1988; Landschulz et al., 1989; Pei and Shih, 1991]. In chicken cell lines, C/EBPα and C/EBPβ are both expressed in myeloid and eosinophilic cells, but not in lymphoid, MEP, thrombocytic, or erythroid cells [Nerlov et al., 1998].

To examine the role of C/EBPs in hematopoiesis, C/EBPα and C/EBPβ were placed downstream of IRES elements in E26 vectors allowing their expression in primary virus transformants [Nerlov et al., 1998]. In these experiments, infection of blastoderms with wild-type E26 vectors yielded 50% myeloid colonies, 50% MEP colonies, and few or no eosinophils. In contrast, both isoforms of C/EBP produced a dramatic increase in eosinophil colonies and a compensatory decrease in myeloid colonies. This shows that C/EBPα and C/EBPβ both induce eosinophil differentiation (fig. 2E).

In order to assess the kinetics of C/EBPβ-induced differentiation, time-course experiments were performed using a hormone-inducible fusion protein (C/EBPβ-ER) [Nerlov et al., 1998]. In these experiments, C/EBPβ activity was induced in MEP cells by administration of β-estradiol for 1–4 days. The hormone was then removed by thorough washing, differentiation was allowed to proceed, and all samples were analyzed after a total of 8 days. Activation of C/EBPβ for 1 day resulted in significant eosinophil commitment with 10–30% of cells expressing EOS47; however, these cells were immature based on their morphology and their lack of granules and eosinophil-peroxidase. A 2-day pulse led to 80% EOS47$^+$ cells, including a high proportion of cytoplasmic granule-containing, peroxidase-positive cells. In addition, 5% of the cells became myelomonocytic (expressing the myeloid-specific MYL51/2 antigen). C/EBPβ expression for 4 days increased the proportion of myeloid cells up to 15–20% of the total. These results demonstrate that, in addition to eosinophil commitment, C/EBPβ induces commitment to the myelomonocytic lineage, although eosinophil commitment was always seen to precede myeloid lineage commitment. The results seen with C/EBPβ (80% of MEPs differentiating into eosinophils after 2 days) contrasts sharply with the results seen for PU.1, where expression in MEPs led to exclusively myeloid differentiation. However, since C/EBPβ also induces myeloid differentiation, there is likely to be some functional overlap with PU.1.

Interestingly, when mutants bearing only the DNA-binding domain and lacking the transactivation domain of C/EBPβ were used, pro-eosinophils were generated [Nerlov et al., 1998]. These cells expressed high levels of the eosinophil-specific EOS47 antigen but, in contrast to full-length C/EBPβ transformants, they lacked expression of mature eosinophil granule proteins. These data suggest that the DNA-binding domain of C/EBPβ alone is sufficient for the commitment of progenitors to the eosinophil lineage, but that the transactivation domain is also required for terminal differentiation into mature eosinophils. Finally, several C/EBP molecules with different transactivation strengths were compared for their ability to induce eosinophil formation in the E26 system and to transactivate the EOS47 promoter in fibroblasts (see below). The results showed a direct correlation: mutants with the weakest transactivation strength on the EOS47 promoter gave the lowest number of eosinophils in the E26 transformation assay, while strong transactivators gave high numbers of eosinophils.

It is noteworthy that in addition to the observed correlation between the strength of the C/EBP transactivation domain and the molecule's ability to generate eosinophils in the E26 system, there was also a correlation with the ability of these molecules to generate myelomonocytic cells [Nerlov et al., 1998]. C/EBP mutants with weaker transactivation domains tended to permit myelomonocytic differentiation while stronger transactivation domains tended to drive exclusively eosinophil differentiation, although a complete absence of the transactivation domain also prevented myelomonocytic differentiation. Consistent with these observations, C/EBPα (which has a much stronger transactivation domain than C/EBPβ) produced exclusively eosinophils in the E26 system. Thus C/EBPβ induces eosinophil and myeloid differentiation while C/EBPα induces eosinophil differentiation only (fig. 2E).

In summary, C/EBP-type transcription factors appear to play a very important role in regulation of maturation and lineage commitment. Highly active members drive eosinophil commitment and terminal eosinophil differentiation. Weaker members drive differentiation along both myelomonocytic and eosinophilic lineages, but are poorer at inducing terminal eosinophil maturation. Finally, the DNA-binding domain of C/EBPβ alone is insufficient for commitment to the myelomonocytic lineage, but it can drive commitment to the eosinophil lineage. Terminal eosinophil differentiation, however, is still dependent on the transactivation domain.

GATA-1, FOG-1, and C/EBPβ: A Complex Interplay Regulates Various Aspects of Hematopoiesis

If FOG-1 modulates GATA-1 activity, then what regulates FOG-1 expression? Since down-regulation of FOG-1 is required for commitment to the eosinophil lineage, and C/EBPβ induces commitment to this lineage, a logical assumption is that C/EBPβ regulates FOG-1 expression. To test this idea, a hormone-inducible form of C/EBPβ was expressed in an E26-transformed MEP cell line [Querfurth et al., 2000]. C/EBP activation by estradiol treatment resulted in a rapid decrease in FOG-1 mRNA levels, followed by increased expression of the eosinophil antigen, EOS47. GATA-1 expression, however, was not altered during this time-frame. Similar results were observed when C/EBPβ-ER fusions were expressed in primary E26 transformants. Treatment of these MEP cells with β-estradiol for 0–3 days, followed by replating in the absence of estradiol showed that levels of FOG-1 mRNA decreased after 1 day, and it was completely down-regulated after a 3-day induction. Eosinophil lineage-commitment was significant after 2 days, and most cells were fully committed within 3 days. These data suggest that FOG-1 must be down-regulated before C/EBPβ-mediated up-regulation of eosinophil-specific gene expression can occur. It is interesting to note that high C/EBP expression can be maintained in the presence of GATA-1 (in eosinophils) or FOG-1 (in FOG-1-expressing myeloblasts), but not in the presence of both (in MEPs). This suggests that C/EBP expression can be suppressed by the combinatorial action of GATA-1 and FOG-1.

In summary, the E26 system has been instrumental in unraveling the complex web of transcriptional control that regulates the maintenance and formation of progenitor-, eosinophil-, and myelomonocytic-type cells. GATA-1 and PU.1 regulate the balance between eosinophil and myelomonocytic differentiation by direct protein-protein interactions leading to mutual repression. FOG-1 modulates this repression and determines whether cells maintain a progenitor phenotype or undergo maturation along the eosinophilic lineage. C/EBP family members are required for eosinophilic differentiation and transcriptionally repress the activity of FOG-1.

The E26 System Can Be Used to Identify Lineage-Specific Genes and Analyze Their Regulation

Because of the robust nature of the E26 system, it has been used in 2 ways to identify novel lineage-specific hematopoietic genes. The first involved the use of temperature-sensitive mutants of the E26 virus to screen for target genes of v-Myb and v-Ets via differential mRNA expression [Kraut et al., 1995; Ness et al., 1989]. The second involved use of various cell types produced in the E26 system to generate monoclonal antibodies and screen for lineage-specific membrane protein expression [McNagny et al., 1992]. In both cases, these screens led to the identification of genes that have proven invaluable for the study of hematopoietic differentiation and, in addition, are beginning to shed light on how hematopoietic cells interact with their in vivo microenvironments. Although it is beyond the scope of this review, it is noteworthy that the well-defined phenotypes and gene expression profiles of E26-transformed cells have also proven extremely valuable in assessing the role of chromatin structure and histone acetylation in stage-specific gene expression [Kontaraki et al., 2000; Litt et al., 2001]. These studies are producing new paradigms for how genes are regulated during hematopoietic development.

Screen for Myb and Ets Target Genes Using Temperature-Sensitive Mutants

Because temperature-sensitive E26 clones can be 'prescreened' for reproducible and high efficiency differentiation, they are an ideal system for identification of genes that are regulated by either Myb or Ets or that change in expression during hematopoietic differentiation [Frampton et al., 1996; Kowenz-Leutz et al., 1997; Kraut et al., 1995; Ness et al., 1989]. Since temperature shift-induced differentiation is reversible for the first 48 h, one simply screens for mRNAs that are up- or down-regulated in cells that are temperature shifted for 24 h (to inactivate either Myb or Ets) and then 'backshifted' to reactivate the respective transcription factor. Inactivation of Myb in such a system led to the cloning of the first Myb-regulated gene, called *mim-1* [Ness et al., 1989]. *Mim-1* contains 3 Myb-binding sites in its promoter region and is expressed in myeloblasts, promyelocytes, and eosinophils, but its role in differentiation and/or transformation is unknown. Although Myb is expressed in early hematopoietic progenitors, Mim-1 is not; this is probably due to the required cooperation of Myb and C/EBPβ for optimal expression [Ness et al., 1993]. Interestingly, cotransfection of C/EBPβ and Myb into heterologous cell types (fibroblasts) led to activation of the endogenous *mim-1* gene suggesting that these 2 factors are sufficient for the expression of at least 1 myeloid gene [Ness et al., 1993].

Similar temperature-sensitive screens for Ets regulated genes led to the cloning of *rem-1*, a recoverin- and visinin-related calcium-binding protein expressed in early hematopoietic cell lines, bone marrow, gut, and neural tissue [Kraut et al., 1995]. Inactivation of Ets DNA-binding activity by temperature shift of Ets^{ts}-transformed MEPs led to a rapid down-regulation of Rem-1, while the subsequent reactivation of Ets by shifting cells back to the permissive temperature resulted in rapid re-expression of Rem-1, even in the absence of de novo protein synthesis. Whether this represents a direct action of Ets on the *rem-1* promoter remains to be shown [Kraut et al., 1995].

In addition to the genes identified directly through the use of temperature-sensitive mutants, several other genes that were suspected to be Myb and Ets targets have been confirmed using this system and IRES-based E26 vectors [Frampton et al., 1996; Kowenz-Leutz et al., 1997]. For example, Frampton et al. have presented elegant data to show that endogenous bcl-2 is directly regulated by Myb in myelomonocytic cells (probably through direct promoter binding) and that the apoptotic death of Myb^{ts} macrophages after temperature shift could be rescued by addition of ectopic bcl-2 in a Myb^{ts}-IRES vector [Frampton et al., 1996]. Surprisingly, temperature-sensitive inactivation of Myb in progenitor cells did not result in apoptosis, suggesting that in early progenitors/thrombocytes other factors are sufficient to maintain bcl-2 expression in the absence of v-Myb.

Screen for Differentially Expressed Genes by Antibody Production

A number of lineage-specific monoclonal antibodies have been produced using the E26 system. This was accomplished by immunizing mice with a pool of E26-transformed multipotent progenitor clones, fusing spleen cells of these mice with a myeloma cell line, and then screening the resulting hybridomas for reactivity with progenitor cells but not with more mature cells [McNagny et al., 1992]. Of the antibodies that were highly reactive against progenitor cell clones, 3 were found to be specific since they had reduced or no activity against differentiated eosinophils and myeloblasts or cell lines representative of other avian hematopoietic lineages. These were named MEP17, MEP21, and MEP26. MEP17 was subsequently shown to recognize the avian homologue of the collagen receptor, VLA-2 or $\alpha_2\beta_1$ integrin [Bradshaw et al., 1995; Mey et al., in press]. Since the MEP17 antibody is a potent inhibitor of collagen-dependent adhesion, it has been an extremely useful tool for dissecting the importance of collagen binding in a variety of cellular systems. MEP26 is a novel 50-kDa homodimeric molecule of unknown function [McNagny et al., 1992]. Finally, MEP21 is a close relative of the mammalian stem cell-specific protein, CD34 and appears to be crucial for normal development in vivo (see below). In addition to the antibodies raised against progenitor cells, a fourth antibody was raised by immunizing with E26-transformed progenitors that had been induced to differentiate into eosinophils [McNagny et al., 1992]. This antibody recognizes a protein called EOS47 that is exquisitely specific for eosinophils (the most specific marker yet identified in any species), and the analysis of the promoter for this gene has provided mechanistic insights into the regulation of gene expression by GATA-1, C/EBP, and FOG-1 [McNagny et al., 1996, 1998]. For the remainder of this review we will focus attention on the function and transcriptional control of EOS47 and MEP21.

EOS47

As mentioned, the EOS47 antigen was first detected as an eosinophil-specific molecule during antibody production of lineage-specific markers using the E26 system [McNagny et al., 1992]. Although it is expressed by several non-hematopoietic tissues, within the hematopoietic system EOS47 is exclusively expressed by eosinophils [Graf et al., 1992; McNagny et al., 1996] (fig. 4A). Subsequent cloning and sequence analysis suggest that EOS47 is an unusual member of the transferrin supergene family of iron-binding proteins [McNagny et al., 1996], and our data suggest that it may have an immune function (see *EOS47 structure and function* section below). With regard to its regulation in eosinophils, we have found that the onset of EOS47 expression correlates well with commitment of stem cells to eosinophil differentiation and precedes the expression of other known eosinophil markers and the formation of eosinophilic granules [McNagny et al., 1992]. Thus, the analysis of its regulatory elements offers a unique opportunity to delineate the control of eosinophil-specific gene expression.

The EOS47 Promoter

RNase protection analysis suggests that *eos47* transcription initiates from a TATA-less promoter approximately 150 bp upstream of the first ATG of the coding sequence [McNagny et al., 1996] (fig. 4B). Transfection of reporter gene constructs carrying approximately 300 bp of sequence upstream from the ATG into myeloblasts, eosinophils, and MEP cells demonstrated that this region likely contains all of the elements required for eosinophil-specific gene expression. When compared to the promoter of the human *eos47* gene (melanotransferrin, see below), conserved binding sites for Myb, Ets, and C/EBP family members were observed in this region. In addition, there was a GATA binding site downstream of the chicken *eos47* major transcription initiation site which was not present in the human gene, although a number of GATA sites are found upstream from the human initiation site [McNagny et al., 1998, and unpublished observations].

Experiments examining the functionality of these putative transcription factor-binding sites suggested that the *eos47* promoter is regulated by a combination of Myb, Ets-1/Fli-1, GATA-1, and C/EBP. In addition, it appears that C/EBPα and Ets-1 interact via protein-protein interactions and cooperate in high affinity binding to the combined binding site in the promoter. Since Ets-1 is expressed in eosinophils as well as MEPs, but C/EBPα and β are not expressed in MEPs, it seems likely that induction of C/EBPs in progenitor cells is critical for eosinophil commitment. Furthermore, the results demonstrated that GATA-1 acts as a modulator of expression by activating transcription at low concentrations and repressing at high concentrations. Thus, the experiments with the EOS47 promoter provide mechanistic data to support the importance of the key transcription factors identified independently in overexpression experiments with E26 [Kulessa et al., 1995; McNagny et al., 1998; Nerlov et al., 1998].

Structure and Potential Function of EOS47

Cloning and sequence analysis of chicken EOS47-encoding cDNAs revealed a 738-amino-acid sequence with striking homology to members of the transferrin family of iron-binding proteins [McNagny et al., 1996]. The highest homology (61.1% identity) was found with the human tumor-associated antigen, melanotransferrin or p97. The high homology, together with the absolute conservation of all 28 cysteine residues, suggests that EOS47 and melanotransferrin are orthologues. EOS47/melanotransferrin is unique among the members of the transferrin supergene family in that alternate splicing gives rise to both membrane-bound and secreted forms of the molecule. Although melanotransferrin has been speculated to function as an alternative to the classic transferrin/transferrin receptor-mediated iron absorption pathway, several observations suggest this may not be the case.

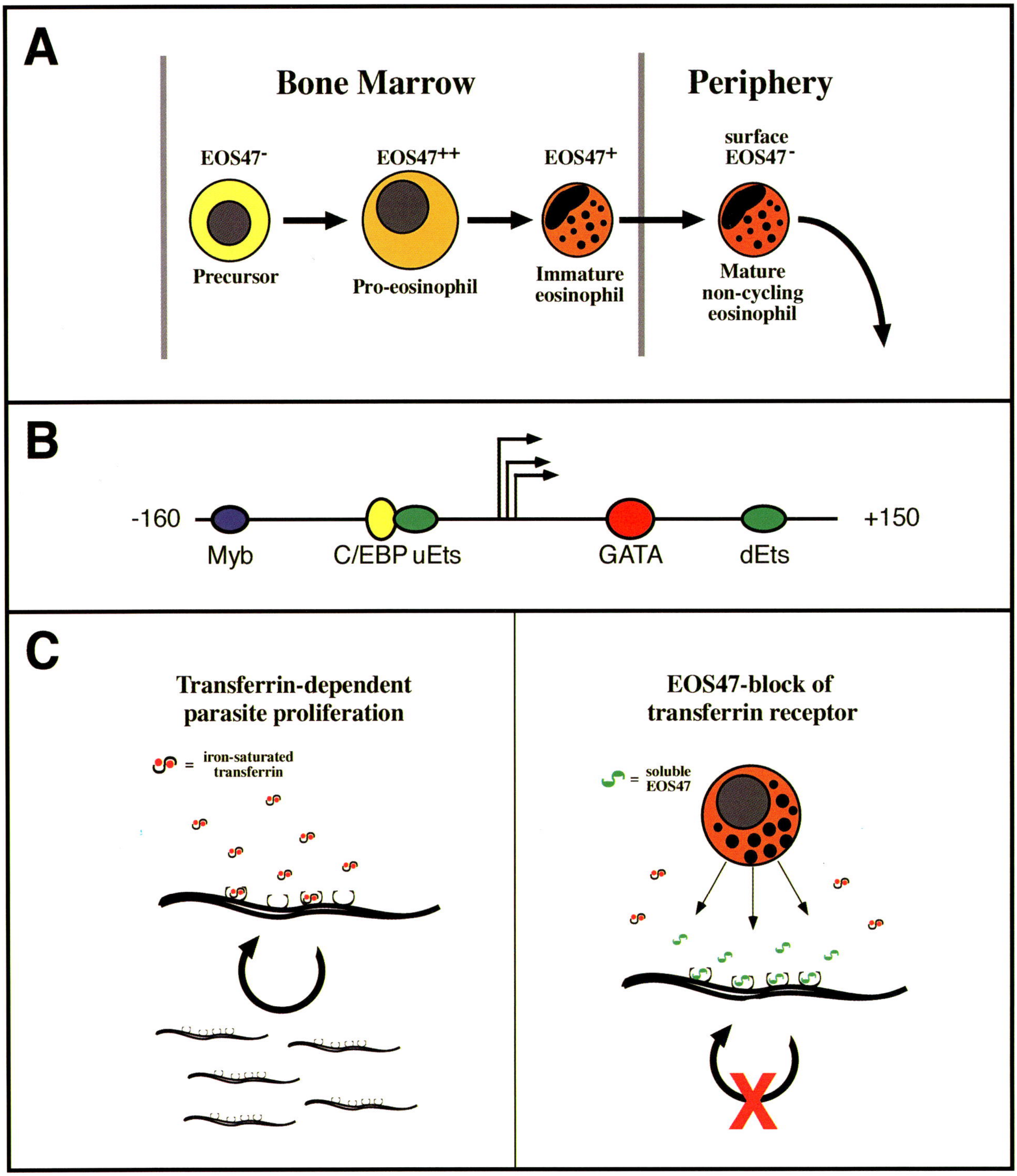

Fig. 4. Promoter, expression pattern, and potential function of the eosinophil gene, EOS47. **A** Expression pattern of EOS47 on normal bone marrow eosinophils and progenitors. **B** Schematic representation of 300-bp *eos47* minimal promoter and transcription factor binding sites. **C** Putative immune function of EOS47 in eosinophils.

Most significantly, human melanotransferrin has point mutations in 1 of the 2 iron-binding repeats that is predicted to lead to a 1,000-fold decreased affinity [McNagny, unpublished modelling study]. Interestingly, chicken EOS47 has similar mutations in both iron-binding domains [McNagny et al., 1996].

Why synthesize an iron-binding protein that is incapable of binding iron? An intriguing hypothesis is that EOS47 serves an innate immune function since the primary function of eosinophils is to fight parasitic infections. Iron is essential for the proliferation of all cells, and parasites scavenge iron from vertebrate transferrin via the expression of parasite transferrin receptors. One possibility is that eosinophils inhibit parasite growth by producing a 'decoy transferrin' that blocks serum transferrin uptake (fig. 4C). Alternatively, it is possible that EOS47 may have a novel (non-iron) ligand. Deletion of EOS47 in mice may clarify its true function.

MEP21

Much of our current work involves the hematopoietic progenitor-specific cell-surface marker, MEP21, which we identified previously using the E26 system. This molecule, also known as podocalyxin, PCLP-1, or thrombomucin, is a 150–165 kDa member of the large family of cell-surface sialomucins [Hara et al., 1999; McNagny et al., 1997; Sassetti et al., 1998]. Genomic organization, mRNA splicing, amino acid sequence, and protein structure all suggest that MEP21 is most closely related to 2 other sialomucins: CD34 and endoglycan (fig. 5). CD34 is the prototypic member of this family and is currently the most widely used marker of hematopoietic stem cells in human [reviewed in Krause et al., 1996]. Each of these molecules has a serine-threonine-proline-rich mucin domain, a disulphide-bonded globular domain, a 'stalk' domain, a transmembrane region, and a conserved cytoplasmic tail of approximately 70 amino acids (fig. 5). The cytoplasmic domain contains regions of linear sequence homology between the family members as well as potential phosphorylation sites for protein kinase C (PKC) and casein kinase II (CKII) and a potential carboxy-terminal docking site for PDZ domain proteins (fig. 5A). These 3 molecules are strikingly similar in exon size and number as well as in the protein domains specified by each exon. In addition, the intronic distances are remarkably similar, and all 3 have similar patterns of alternative splicing leading to production of proteins with truncated cytoplasmic tails (fig. 5B, C). Finally, all 3 members of this family are co-expressed by hematopoietic precursors and subsets of vascular endothelial cells. In addition, MEP21/podocalyxin is also expressed by platelets, podocytes in the kidney, and mesothelial cells [McNagny et al., 1997], while endoglycan has a broader expression pattern and is present on some mature blood cells, pericytes and smooth muscle, and a subset of neuronal cells [Sassetti et al., 2000; McNagny and Merkens, unpublished].

Transcriptional control of the human *cd34* locus appears to be remarkably complex [Radomska et al., 1998; Yamaguchia et al., 1997]. Attempts to define the minimal regulatory sequences required for stem cell-specific expression suggest that over 40 kb (including upstream, downstream, and intronic sequences) are necessary. Preliminary experiments with the *mep21* locus suggest that it may be regulated in an equally complex manner [Frampton and Doyonnas, unpublished].

Incredibly, although there are over 7,000 CD34-related publications to date, the function of this molecule is essentially unknown. In order to elucidate the function of podocalyxin, and potentially its related family members, we have performed both gain of function and loss of function experiments.

Function of MEP21 and CD34-Type Proteins: Overexpression Experiments

In the E26 system MEP21 is expressed by MEPs, but expression is lost upon lineage commitment and differentiation. We therefore used this system to analyze the role of MEP21 in hematopoiesis by gain of function experiments. We transformed myeloblasts with Myb^{ts} (the E26 variant with a temperature sensitive mutation in the DNA-binding domain of Myb) and expressed MEP21 in these cells using an IRES element downstream of the oncogene. When these cells were induced to differentiate into mature myeloid cells by temperature shift, they acquired morphological features of macrophages but, in contrast to MEP21 non-expressing clones, these clones displayed a reduced capacity for adhesion to plastic [Nielsen and McNagny, unpublished]. This is consistent with the fact that mucin-type molecules can block adhesion by steric hindrance (resulting from extensive O-linked glycosylations) or charge repulsion (due to the presence of sialic acid residues [Wesseling et al., 1996]). Previously, CD34 has been ectopically expressed in murine cell lines capable of macrophage differentiation and found to 'block' maturation of these cells [Fackler et al., 1995]. It is noteworthy, however, that the principal features used to measure myeloid maturation in these assays were adhesion and phagocytosis. It is therefore likely that

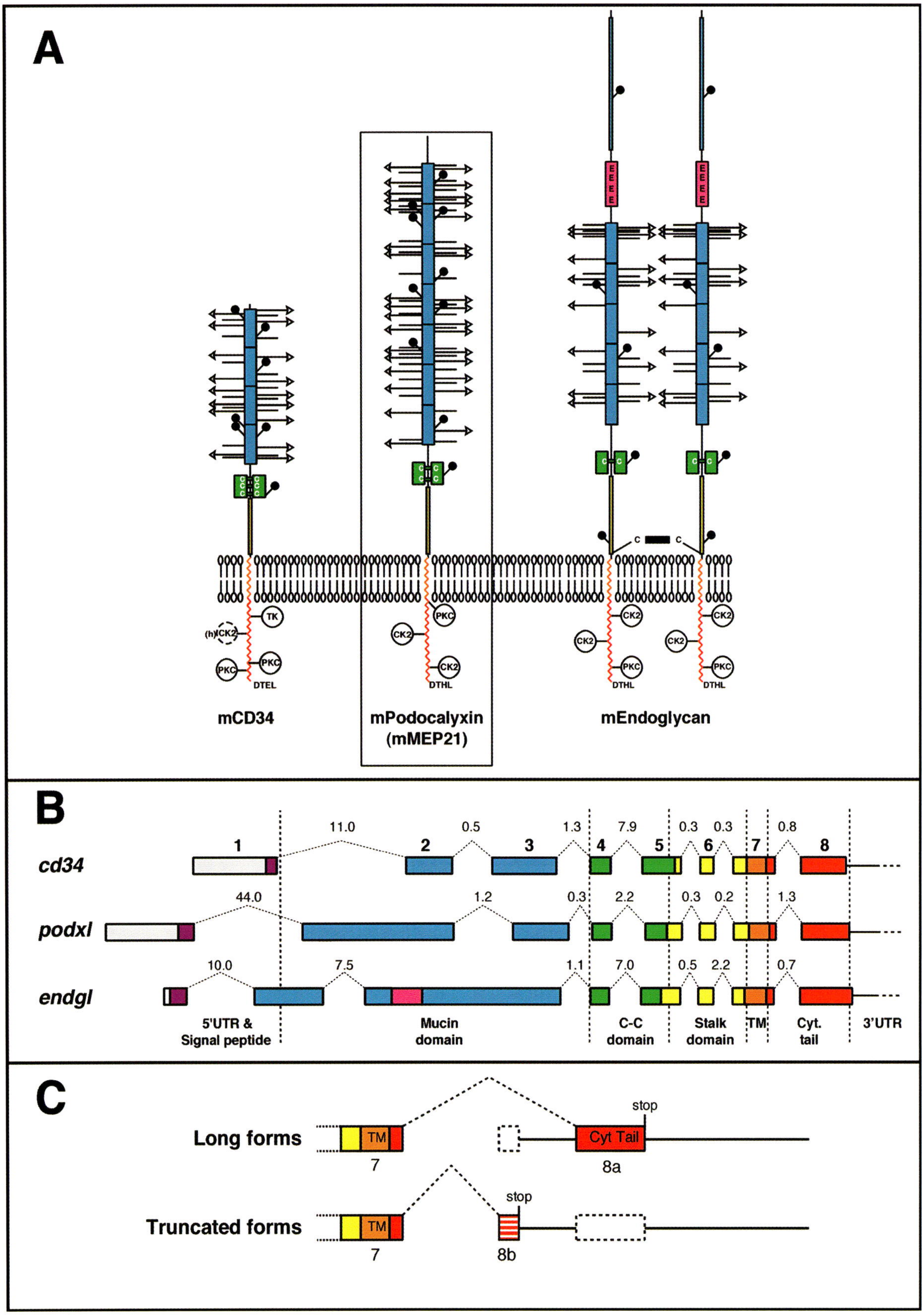
A
mCD34
mPodocalyxin
(mMEP21)
mEndoglycan
TK
(h)CK2
PKC
DTEL
CK2
DTHL
B
cd34
podxl
endgl
11.0
0.5
1.3
7.9
0.3
0.8
44.0
1.2
2.2
0.2
10.0
7.5
1.1
7.0
0.5
0.7
5'UTR &
Signal peptide
Mucin
domain
C-C
domain
Stalk
domain
TM
Cyt.
tail
3'UTR
C
Long forms
Truncated forms
TM
Cyt Tail
stop
7
8a
8b

CD34, too, exerts an anti-adhesive effect in overexpression experiments. Finally, Takeda et al. [2000] also analyzed the anti-adhesive properties of MEP21/podocalyxin by expression in Chinese hamster ovary (CHO)-K1 and Madin-Darby canine kidney (MDCK) cells. Using an in vitro cell aggregation assay, they observed that increasing levels of MEP21 expression led to decreased aggregation of CHO-K1 cells. In addition, expression of MEP21 by MDCK monolayers grown in trans-well cultures resulted in the disruption of intercellular tight junctions and a decrease in transepithelial electrically resistant monolayers. These results are therefore consistent with an anti-adhesive function.

Role of MEP21 and CD34-Type Proteins during Development: Loss of Function Experiments

In mice, targeted deletion of the *cd34* gene was surprisingly uninformative in that null mice developed normally and exhibited only very subtle defects in hematopoietic maturation and function [Cheng et al., 1996; Suzuki et al., 1996]. This is likely due to functional compensation by other members of the family. To test this hypothesis, we generated MEP21-deficient mice with the intent of crossing them to CD34-deficient mice [Doyonnas et al., 2001].

Fig. 5. Structure of CD34-related proteins and genomic organization of *cd34*-related genes. **A** Schematic representation of the structure of murine CD34, podocalyxin/MEP21, and endoglycan based on predicted protein sequences. Blue boxes = mucin domains; green boxes = cysteine-rich domains; circles = potential N-linked carbohydrates; horizontal bars with or without arrows = potential O-linked carbohydrates; arrows = potential sialic acid motifs on O-linked carbohydrates; PKC, TK and CK2 = potential phosphorylation sites; DTHL or DTEL = potential PDZ domain docking sites. **B** Schematic showing the genomic organization of human *cd34*, *podxl*, and *endgl* genes based on sequence contigs identified in the NCBI human sequence database. The location of the coding sequences for the signal peptide (purple), mucin domain (blue), cysteine rich domain (C-C, green), transmembrane domain (TM, orange), and cytoplasmic tail (Cyt. tail, red) are indicated. Numbers indicate intron size in kilobase pairs. **C** Alternative splicing mechanism leading to the production of full-length or truncated forms of CD34-family members lacking the cytoplasmic domain. In all 3 family members alternative splicing between exons 7 and 8 gives rise to either a cytoplasmic tail-encoding form (short transcripts) or transcripts with a premature stop codon deleting most of the cytoplasmic tail (long transcripts).

In contrast to the mild phenotype seen in CD34 knockouts, MEP21 deficiency itself results in perinatal lethality in mice due to profound defects in kidney development. In the MEP21 knockout mice the podocytes do not form foot processes and the slit diaphragms are absent. Instead, podocytes retain tight junctions and adherens junctions similar to those found in immature glomeruli [Doyonnas et al., 2001]. This results in a profound absence of urine production, elevated blood pressure, and perinatal lethality. Persistence of junctional complexes and lack of foot processes/pseudopods in the absence of MEP21 supports the hypothesis that MEP21 interacts with the actin cytoskeleton and leads to relocalization of junctional proteins.

Interestingly, MEP21-deficient mice display no defects in tissues where MEP21 and CD34 are co-expressed (hematopoietic and vascular cells) suggesting that CD34 functionally compensates for podocalyxin loss [Krause et al., 1996; McNagny et al., 1997]. This is supported by the observation that the vascular endothelial cells in MEP21 knockout mice express 2–4 fold higher levels of CD34 than do wild-type mice [Doyonnas et al., 2001].

CD34, MEP21, and Endoglycan: Adhesion or Anti-Adhesion Molecules?

As mentioned above, both CD34 and MEP21 are expressed by vascular endothelia [McNagny et al., 1997; Sassetti et al., 1998]. Previous experiments in a number of laboratories have demonstrated a role for CD34 and MEP21/podocalyxin in the adhesion of leukocytes to high endothelial venules (HEV) [Baumhueter et al., 1993; Sassetti et al., 1998] (fig. 6). These leukocytes express L-selectin, which binds to specific glycoforms of CD34 and MEP21 expressed by HEV. HEV are specialized post-capillary venules present in secondary lymphoid organs and at sites of chronic inflammation. Binding, and the subsequent selectin-dependent 'rolling' of leukocytes on endothelia, represents the initiating step in leukocyte extravasation from the blood and into secondary lymphoid tissues. L-selectin binding, however, is critically dependent on HEV-specific glycosylation of CD34 and MEP21; the appropriate modifications for binding have not been found on most vasculature and hematopoietic stem cells. Thus, it is likely that these molecules serve a different function on these cells.

In contrast, our own data in MEP21 knockout mice and in overexpression experiments suggest all defects can be ascribed to loss of anti-adhesive properties of these molecules. Which hypothesis is correct? An intriguing possibility is that these molecules fulfill both adhesive and

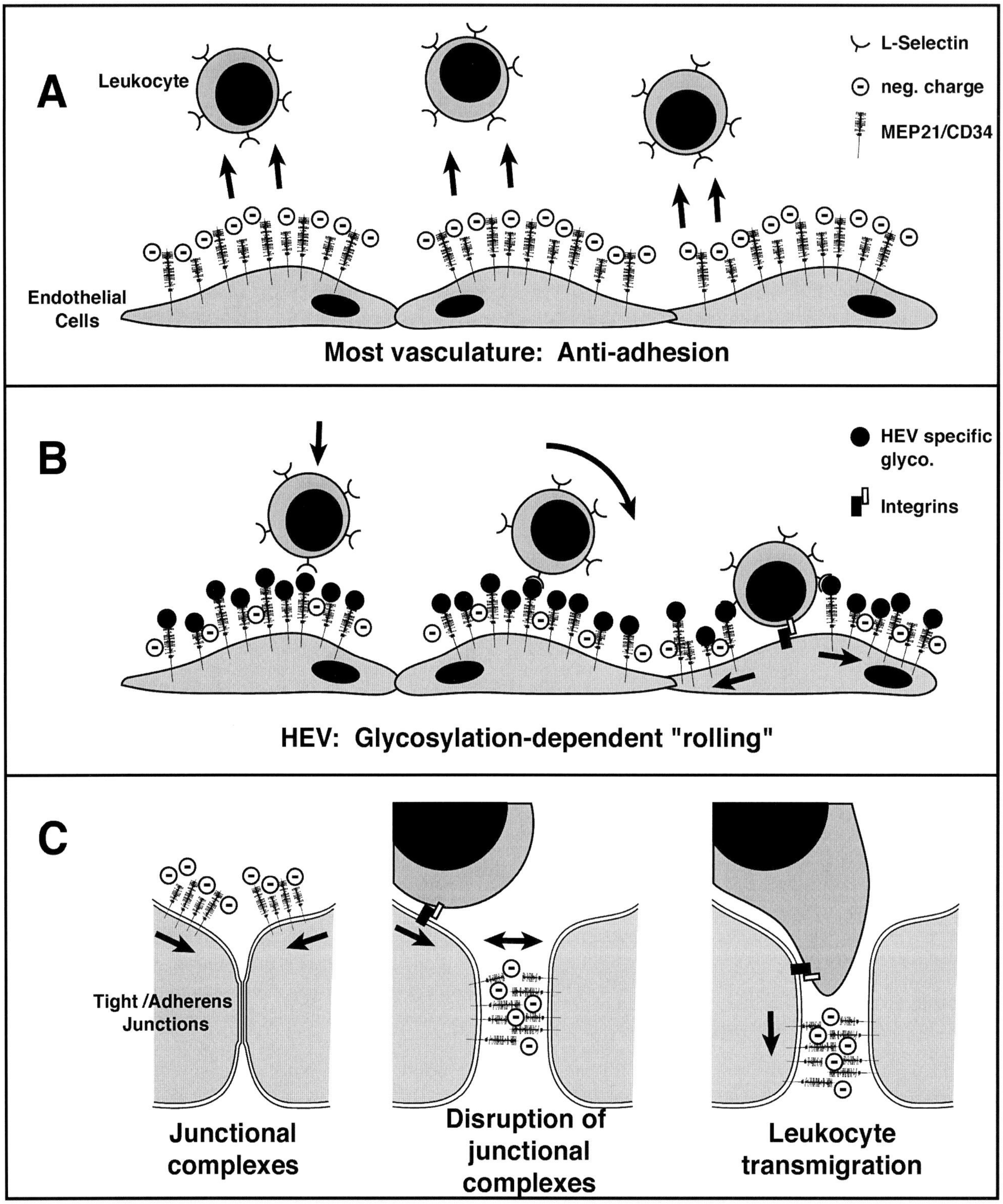

Fig. 6. Model for CD34 family function on vascular endothelia. **A** On most vessels, the high density of negative charges (due to sulfation and sialic acid residues) on CD34-type proteins provides a barrier to cell adhesion. **B** On high endothelial venules (HEV), tissue specific-glycosylation of CD34-type proteins provides a ligand for L-selectin on migrating leukocytes. This initiates leukocyte 'rolling', a first step in leukocyte extravasation. **C** Activation-dependent migration of CD34 family members between endothelial cells results in disruption of cell-cell junctional complexes and facilitates leukocyte extravasation.

anti-adhesive functions. Globally, CD34 and MEP21 may function to reduce intercellular adhesion (fig. 6A). In the special case of HEV, CD34 and MEP21 may first act as adhesive tethers for leukocytes via HEV-specific glycosylation (fig. 6B). Subsequently, movement of these molecules to the junctions between cells could directly regulate the disruption of the endothelial cell junctions and facilitate transmigration through the endothelium (fig. 6C). This is consistent with previous observations showing that activation of endothelial cells leads to movement of CD34 to intercellular junctions [Delia et al., 1993; Fackler et al., 1990].

Concluding Remarks

In this review, we have demonstrated how the E26 system can be used to gain insights into the complex regulation of hematopoiesis by transcription factors such as GATA-1, FOG-1, PU.1, and C/EBP. Furthermore, this system has revealed an amazing plasticity in fate determination of hematopoietic cells, which was unthinkable only a few years ago. It has become increasingly apparent that by shifting the balance of transcription factors one can induce trans- and retro-differentiation in hematopoietic cells [Orkin, 2000; Sieweke and Graf, 1998]. A challenge for the future is to identify when and if such changes occur in disease states (neoplasia or anemia) or whether it is possible to manipulate the transcription factor balance in normal cells. In either case, the E26 system offers a simple and effective model for prescreening the potential outcomes of perturbation in the transcription factor web. In addition, we have shown that this system provides a tool for identification of lineage-specific genes that have not been found in mammalian systems.

Acknowledgements

The authors wish to thank Annette Schallhorn, Thomas Graf and Helen Merkens for critical evaluation of this manuscript and Shierley Chelliah, Helen Merkens, and Erin Drew for technical assistance with MEP21 experiments. This work was funded by Canadian MRC grant #MT-15477 (KMM), a grant from the Heart and Stroke Foundation of British Columbia, and a fellowship from the National Sciences and Engineering Research Council of Canada (JSN). KMM is a Canadian Medical Research Council Scholar.

References

Baumhueter, S., M. Singer, W. Henzel, S. Hemmerich, M. Renz, S. Rosen, L.A. Lasky (1993) Binding of L-selectin to the vascular sialomucin CD34. Science *262:* 436.

Beug, H., A. Leutz, P. Kahn, T. Graf (1984) Ts mutants of E26 leukemia virus allow transformed myeloblasts, but not erythroblasts or fibroblasts to differentiate at the nonpermissive temperature. Cell *39:* 579–588.

Bradshaw, A., K. McNagny, D. Gervin, G. Cann, T. Graf, D. Clegg (1995) Integrin a_2b_1 mediates interactions between developing embryonic retinal cells and collagen. Development *121:* 3593–3602.

Cheng, J., S. Baumhueter, G. Cacalano, K. Carver-Moore, H. Thibodeaux, R. Thomas, H.E. Broxmeyer, S. Cooper, N. Hague, M. Moore, L.A. Lasky (1996) Hematopoietic defects in mice lacking the sialomucin CD34. Blood *87:* 479–490.

Crispino, J.D., M.B. Lodish, J.P. MacKay, S.H. Orkin (1999) Use of altered specificity mutants to probe a specific protein-protein interaction in differentiation: the GATA-1:FOG complex. Mol Cell *3:* 219–228.

DeKoter, R.P., H. Singh (2000) Regulation of B lymphocyte and macrophage development by graded expression of PU.1. Science *288:* 1439–1441.

Delia, D., M.G. Lampugnani, M. Resnati, E. Dejana, A. Aiello, E. Fontanella, D. Soligo, M.A. Pierotti, M.F. Greaves (1993) CD34 expression is regulated reciprocally with adhesion molecules in vascular endothelial cells in vitro. Blood *81:* 1001–1008.

Doyonnas, R., D.B. Kershaw, C. Duhme, H. Merkens, S. Chelliah, T. Graf, K.M. McNagny (2001) Anuria, omphalocele, and perinatal lethality in mice lacking the CD34-related protein podocalyxin. J Exp Med *194:* 13–27.

Evans, T., G. Felsenfeld (1989) The erythroid-specific transcription factor Eryf1: a new finger protein. Cell *58:* 877–885.

Fackler, M.J., C.I. Civin, D.R. Sutherland, M.A. Baker, W.S. May (1990) Activated protein kinase C directly phosphorylates the CD34 antigen on hematopoietic cells. J Biol Chem *265:* 11056–11061.

Fackler, M.J., D.S. Krause, O.M. Smith, C.I. Civin, W.S. May (1995) Full-length but not truncated CD34 inhibits hematopoietic cell differentiation of M1 cells. Blood *85:* 3040–3047.

Fox, A.H., K. Kowalski, G.F. King, J.P. Mackay, M. Crossley (1998) Key residues characteristic of GATA N-fingers are recognized by FOG. J Biol Chem *273:* 33595–33603.

Fox, A.H., C. Liew, M. Holmes, K. Kowalski, J. Mackay, M. Crossley (1999) Transcriptional cofactors of the FOG family interact with GATA proteins by means of multiple zinc fingers. EMBO J *18:* 2812–2822.

Frampton, J., K.M. McNagny, M. Sieweke, A. Philip, G. Smith, T. Graf (1995) v-Myb DNA binding is required to block thrombocytic differentiation of Myb-Ets transformed multipotent haematopoietic progenitors. EMBO J *14:* 2866–2875.

Frampton, J., T. Ramqvist, T. Graf (1996) v-Myb of E26 leukemia virus up-regulates bcl-2 and suppresses apoptosis in myeloid cells. Genes Dev *10:* 2720–2731.

Freytag, S.O., T.J. Geddes (1992) Reciprocal regulation of adipogenesis by Myc and C/EBP alpha. Science *256:* 379–382.

Frykberg, L., T. Metz, G. Brady, M. Introna, H. Beug, B. Vennstrom, T. Graf (1988) A point mutation in the DNA binding domain of the v-myb oncogene of E26 virus confers temperature sensitivity for transformation of myelomonocytic cells. Oncogene Res *3:* 313–322.

Golay, J., M. Introna, T. Graf (1988) A single point mutation in the v-ets oncogene affects both erythroid and myelomonocytic cell differentiation. Cell *55:* 1147–1158.

Graf, T. (2000) Transcription factors that induce the commitment of multipotent progenitors: lessons from the MEP system. Hematopoiesis: A Developmental Approach. Oxford University Press.
Graf, T., K. McNagny, G. Brady, J. Frampton (1992) Chicken 'erythroid' cells transformed by the gag-myb-ets-encoding E26 leukemia virus are multipotent. Cell *70:* 201–213.
Hara, T., Y. Nakano, M. Tanaka, K. Tamura, T. Sekiguchi, K. Minehata, N.G. Copeland, N.A. Jenkins, M. Okabe, H. Kogo, Y. Mukouyama, A. Miyajima (1999) Identification of podocalyxin-like protein 1 as a novel cell surface marker for hemangioblasts in the murine aorta-gonad-mesonephros region. Immunity *11:* 567–578.
Heyworth, C., K. Gale, M. Dexter, G. May, T. Enver (1999) A GATA-2/estrogen receptor chimera functions as a ligand-dependent negative regulator of self-renewal. Genes Dev *13:* 1847–1860.
Kelly, L.M., U. Englmeier, I. Lafon, M.H. Sieweke, T. Graf (2000) MafB is an inducer of monocytic differentiation. EMBO J *19:* 1987–1997.
Klemsz, M.J., S.R. McKercher, A. Celada, C. Van Beveren, R.A. Maki (1990) The macrophage and B cell-specific transcription factor PU.1 is related to the ets oncogene. Cell *61:* 113–124.
Kontaraki, J., H.H. Chen, A. Riggs, C. Bonifer (2000) Chromatin fine structure profiles for a developmentally regulated gene: reorganization of the lysozyme locus before trans-activator binding and gene expression. Genes Dev *14:* 2106–2122.
Kowenz-Leutz, E., P. Herr, K. Niss, A. Leutz (1997) The homeobox gene GBX2, a target of the myb oncogene, mediates autocrine growth and monocyte differentiation. Cell *91:* 185–195.
Kowenz-Leutz, E., G. Twamley, S. Ansieau, A. Leutz (1994) Novel mechanism of C/EBP beta (NF-M) transcriptional control: activation through derepression. Genes Dev *8:* 2781–2791.
Krause, D.S., M.J. Fackler, C.I. Civin, W.S. May (1996) CD34: structure, biology, and clinical utility [see comments]. Blood *87:* 1–13.
Kraut, N., J. Frampton, T. Graf (1995) Rem-1, a putative direct target gene of the Myb-Ets fusion oncoprotein in haematopoietic progenitors, is a member of the recoverin family. Oncogene *10:* 1027–1036.
Kraut, N., J. Frampton, K.M. McNagny, T. Graf (1994) A functional Ets DNA-binding domain is required to maintain multipotency of hematopoietic progenitors transformed by Myb-Ets. Genes Dev *8:* 33–44.
Kulessa, H., J. Frampton, T. Graf (1995) GATA-1 reprograms avian myelomonocytic cell lines into eosinophils, thromboblasts, and erythroblasts. Genes Dev *9:* 1250–1262.
Landschulz, W.H., P.F. Johnson, S.L. McKnight (1988) The leucine zipper: a hypothetical structure common to a new class of DNA binding proteins. Science *240:* 1759–1764.
Landschulz, W.H., P.F. Johnson, S.L. McKnight (1989) The DNA binding domain of the rat liver nuclear protein C/EBP is bipartite. Science *243:* 1681–1688.
Lemarchandel, V., J. Ghysdael, V. Mignotte, C. Rahuel, P.H. Romeo (1993) GATA and Ets cis-acting sequences mediate megakaryocyte-specific expression. Mol Cell Biol *13:* 668–676.
Leprince, D., A. Gegonne, J. Coll, C. de Taisne, A. Schneeberger, C. Lagrou, D. Stehelin (1983) A putative second cell-derived oncogene of the avian leukemia retrovirus E26. Nature *306:* 395–397.
Litt, M.D., M. Simpson, F. Recillas-Targa, M.N. Prioleau, G. Felsenfeld (2001) Transitions in histone acetylation reveal boundaries of three separately regulated neighboring loci. Embo J *20:* 2224–2235.
Martin, D.I., L.I. Zon, G. Mutter, S.H. Orkin (1990) Expression of an erythroid transcription factor in megakaryocytic and mast cell lineages. Nature *344:* 444–447.
McKercher, S.R., B.E. Torbett, K.L. Anderson, G.W. Henkel, D.J. Vestal, H. Baribault, M. Klemsz, A.J. Feeney, G.E. Wu, C.J. Paige, R.A. Maki (1996) Targeted disruption of the PU.1 gene results in multiple hematopoietic abnormalities. EMBO J *15:* 5647–5658.
McNagny, K.M., T. Graf (1996) Acute avian leukemia viruses as tools to study hematopoietic differentiation. Current Topics in Microbiology and Immunology. Springer-Verlag, Heidelberg, vol 212, pp 143–162.
McNagny, K.M., T. Graf (1997) Production and analysis of retrovirus-transformed multipotent hematopoietic progenitors. Immunology Methods Manual. Academic Press, pp 2185–2197.
McNagny, K.M., F. Lim, S. Grieser, T. Graf (1992) Cell surface proteins of chicken hematopoietic progenitors, thrombocytes and eosinophils detected by novel monoclonal antibodies. Leukemia *6:* 975–984.
McNagny, K.M., I. Pettersson, F. Rossi, I. Flamme, A. Shevchenko, M. Mann, T. Graf (1997) Thrombomucin, a novel cell surface protein that defines thrombocytes and multipotent hematopoietic progenitors. J Cell Biol *138:* 1395–1407.
McNagny, K.M., F. Rossi, G. Smith, T. Graf (1996) The eosinophil-specific cell surface antigen, EOS47, is a chicken homologue of the oncofetal antigen melanotransferrin. Blood *87:* 1343–1352.
McNagny, K.M., M.H. Sieweke, G. Doderlein, T. Graf, C. Nerlov (1998) Regulation of eosinophil-specific gene expression by a C/EBP-Ets complex and GATA-1. EMBO J *17:* 3669–3680.
Metz, T., T. Graf (1991a) Fusion of the nuclear oncoproteins v-myb and v-ets is required for the leukemogenicity of E26 virus. Cell *66:* 95–105.
Metz, T., T. Graf (1991b) v-myb and v-ets transform chicken erythroid cells and cooperate both in trans and in cis to induce distinct differentiation phenotypes. Genes Dev *5:* 369–380.
Mey, M., O. Gandrillon, K.M. McNagny, D.O. Clegg, J. Samarut (in press) The v-erbA oncogene blocks expression of α_2/β_1 integrin, a normal inhibitor of erythroid progenitor proliferation. Oncogene.
Nerlov, C., T. Graf (1998) PU.1 induces myeloid lineage commitment in multipotent hematopoietic progenitors. Genes Dev *12:* 2403–2412.
Nerlov, C., K.M. McNagny, G. Doderlein, E. Kowenz-Leutz, T. Graf (1998) Distinct C/EBP functions are required for eosinophil lineage commitment and maturation. Genes Dev *12:* 2413–2423.
Nerlov, C., E. Querfurth, H. Kulessa, T. Graf (2000) GATA-1 interacts with the myeloid PU.1 transcription factor and represses PU.1-dependent transcription. Blood *95:* 2543–2551.
Ness, S.A., E. Kowenz-Leutz, T. Casini, T. Graf, A. Leutz (1993) Myb and NF-M: combinatorial activators of myeloid genes in heterologous cell types. Genes Dev *7:* 749–759.
Ness, S.A., Å. Marknell, T. Graf (1989) The v-myb oncogene product binds to and activates the promyelocyte-specific mim-1 gene. Cell *59:* 1115–1125.
Nunn, M., P.H. Seeburg, C. Moscovici, P.H. Duesberg (1983) Tripartite structure of the avian erythroblastosis virus E26 transforming gene. Nature *306:* 391–395.
Orkin, S.H. (2000) Diversification of haematopoietic stem cells to specific lineages. Nat Rev Genet *1:* 57–64.
Pei, D.Q., C.H. Shih (1991) An 'attenuator domain' is sandwiched by two distinct transactivation domains in the transcription factor C/EBP. Mol Cell Biol *11:* 1480–1487.
Querfurth, E., M. Schuster, H. Kulessa, J.D. Crispino, G. Doderlein, S.H. Orkin, T. Graf, C. Nerlov (2000) Antagonism between C/EBPbeta and FOG in eosinophil lineage commitment of multipotent hematopoietic progenitors. Genes Dev *14:* 2515–2525.
Radomska, H.S., A.B. Satterthwaite, T.C. Burn, I.A. Oliff, C.S. Huettner, D.G. Tenen (1998) Multiple control elements are required for expression of the human CD34 gene. Gene *222:* 305–318.
Rekhtman, N., F. Radparvar, T. Evans, A.I. Skoultchi (1999) Direct interaction of hematopoietic transcription factors PU.1 and GATA-1: functional antagonism in erythroid cells. Genes Dev *13:* 1398–1411.
Romeo, P.H., M.H. Prandini, V. Joulin, V. Mignotte, M. Prenant, W. Vainchenker, G. Marguerie, G. Uzan (1990) Megakaryocytic and erythrocytic lineages share specific transcription factors. Nature *344:* 447–449.
Rossi, F., K. McNagny, C. Logie, F. Stewart, T. Graf (1996a) Excision of Ets by an inducible site-specific recombinase causes differentiation of Myb-Ets-transformed hematopoietic progenitors. Current Biol *6:* 866–872.
Rossi, F., K. McNagny, G. Smith, J. Frampton, T. Graf (1996b) Lineage commitment of transformed hematopoietic progenitors is determined by the level of PKC activity. EMBO J *15:* 1894–1901.
Sassetti, C., K. Tangemann, M.S. Singer, D.B. Kershaw, S.D. Rosen (1998) Identification of podocalyxin-like protein as a high endothelial venule ligand for L-selectin: parallels to CD34. J Exp Med *187:* 1965–1975.

Sassetti, C., A. Van Zante, S.D. Rosen (2000) Identification of endoglycan, a member of the CD34/podocalyxin family of sialomucins. J Biol Chem *275:* 9001–9010.
Scott, E.W., M.C. Simon, J. Anastasi, H. Singh (1994) Requirement of transcription factor PU.1 in the development of multiple hematopoietic lineages. Science *265:* 1573–1577.
Shivdasani, R., S. Orkin (1996) The transcriptional control of hematopoiesis. Blood *87:* 4025–4039.
Sieweke, M.H., T. Graf (1998) A transcription factor party during blood cell differentiation. Curr Opin Genet Dev *8:* 545–551.
Suzuki, A., D.P. Andrew, J.A. Gonzalo, M. Fukumoto, J. Spellberg, M. Hashiyama, H. Takimoto, N. Gerwin, I. Webb, G. Molineux, R. Amakawa, Y. Tada, A. Wakeham, J. Brown, I. McNiece, K. Ley, E.C. Butcher, T. Suda, J.C. Gutierrez-Ramos, T.W. Mak (1996) CD34-deficient mice have reduced eosinophil accumulation after allergen exposure and show a novel crossreactive 90-kD protein. Blood *87:* 3550–3562.
Takeda, T., W.Y. Go, R.A. Orlando, M.G. Farquhar (2000) Expression of podocalyxin inhibits cell-cell adhesion and modifies junctional properties in madin-darby canine kidney cells [In Process Citation]. Mol Biol Cell *11:* 3219–3232.
Tsai, S.F., D.I. Martin, L.I. Zon, A.D. D'Andrea, G.G. Wong, S.H. Orkin (1989) Cloning of cDNA for the major DNA-binding protein of the erythroid lineage through expression in mammalian cells. Nature *339:* 446–451.
Tsang, A.P., Y. Fujiwara, D.B. Hom, S.H. Orkin (1998) Failure of megakaryopoiesis and arrested erythropoiesis in mice lacking the GATA-1 transcriptional cofactor FOG. Genes Dev *12:* 1176–1188.
Tsang, A.P., J.E. Visvader, C.A. Turner, Y. Fujiwara, C. Yu, M.J. Weiss, M. Crossley, S.H. Orkin (1997) FOG, a multitype zinc finger protein, acts as a cofactor for transcription factor GATA-1 in erythroid and megakaryocytic differentiation. Cell *90:* 109–119.
Umek, R.M., A.D. Friedman, S.L. McKnight (1991) CCAAT-enhancer binding protein: A component of a differentiation switch. Science *251:* 288–292.
Wesseling, J., S.W. van der Valk, J. Hilkens (1996) A mechanism for inhibition of E-cadherin-mediated cell-cell adhesion by the membrane-associated mucin episialin/MUC1. Mol Biol Cell *7:* 565–577.
Yamaguchi, Y., S.J. Ackerman, N. Minegishi, M. Takiguchi, M. Yamamoto, T. Suda (1998) Mechanisms of transcription in eosinophils: GATA-1, but not GATA-2, transactivates the promoter of the eosinophil granule major basic protein gene. Blood *91:* 3447–3458.
Yamaguchi, Y., H. Nishio, K. Kishi, S.J. Ackerman, T. Suda (1999) C/EBPbeta and GATA-1 synergistically regulate activity of the eosinophil granule major basic protein promoter: implication for C/EBPbeta activity in eosinophil gene expression. Blood *94:* 1429–1439.
Yamaguchia, Y., D.G. Tenen, T. Suda (1997) Functional characterization of the promoter for the gene encoding murine CD34. Biochim Biophys Acta *1350:* 141–146.
Zhang, P., G. Behre, J. Pan, A. Iwama, N. Wara-Aswapati, H.S. Radomska, P.E. Auron, D.G. Tenen, Z. Sun (1999) Negative cross-talk between hematopoietic regulators: GATA proteins repress PU.1. Proc Natl Acad Sci USA *96:* 8705–8710.
Zhang, P., X. Zhang, A. Iwama, C. Yu, K.A. Smith, B.U. Mueller, S. Narravula, B.E. Torbett, S.H. Orkin, D.G. Tenen (2000) PU.1 inhibits GATA-1 function and erythroid differentiation by blocking GATA-1 DNA binding. Blood *96:* 2641–2548.
Zon, L.I., M.F. Gurish, R.L. Stevens, C. Mather, D.S. Reynolds, K.F. Austen, S.H Orkin (1991) GATA-binding transcription factors in mast cells regulate the promoter of the mast cell carboxypeptidase A gene. J Biol Chem *266:* 22948–22953.

Cells Tissues Organs

Cells Tissues Organs 2002;171:64–76

Neural Stem Cells: Plasticity and Their Transdifferentiation Potential

Angelo Vescovi Angela Gritti Giulio Cossu Rossella Galli

Stem Cell Research Institute, DIBIT H. San Raffaele, Milan, Italy

Key Words

Neural stem cells · Adult CNS · Neurogenesis · Neural precursors · Niche

Abstract

The presence of resident stem cells in adult tissues is of fundamental importance for the maintenance of their structural and functional integrity. In fact, throughout life, somatic stem cells attend to the critical function of substituting terminally differentiated cells lost to physiological turnover, injury or disease. Thence, one of the basic dogmata in tissue biology holds that the differentiation potential of an adult stem cell is restricted to the generation of the mature cell lineages found in the tissue to which the stem cell belongs. A plethora of recent evidences from many groups, including ours, is now providing evidence that adult stem cells may possess a broader differentiation repertoire than expected and that their fate potential may not be as tissue specific as once thought. The initial example of an unforeseen, trans-germ layer plasticity – that seems now to emerge as a prototypic functional trait of various somatic stem cells of different origin – has come from the reported awakening of a latent hemopoietic developmental capacity in stem cells isolated from the adult mammalian brain following their transplantation into sub-lethally irradiated mice. More recently, it has been shown that adult neural stem cells can differentiate into a wide array of bodily cells of different origin when injected into the blastocyst and into myogenic cells when transplanted into the adult regenerating skeletal muscle. Moreover, bone marrow stem cells can now give rise to skeletal muscle, hepatic and brain cells, whereas muscle precursors can generate blood cells. In this article, we review some of the basic

Abbreviations used in this paper

ANSCs	adult neural stem cells
BDNF	brain-derived neurotrophic factor
BMPs	bone morphogenetic proteins
CNS	central nervous system
CNTF	ciliary neurotrophic factor
EGF	epidermal growth factor
ES cells	embryonic stem cells
FGFs	fibroblasts growth factors
GFs	growth factors
IL-1	interleukin-1
LIF	leukemia inhibitory factor
OB	olfactory bulb
PDGF	platelet-derived growth factor
PSA-NCAM	polysialylated form of neural cell adhesion molecule
SCs	stem cells
SVZ	subventricular zone
TGFα	transforming growth factor α
VZ	ventricular zone
GFAP	glial fibrillary acidic protein

KARGER

Fax + 41 61 306 12 34
E-Mail karger@karger.ch
www.karger.com

1422–6405/02/1711–0064$18.50/0

Accessible online at:
www.karger.com/journals/cto

Angelo Vescovi
Stem Cell Research Institute
DIBIT H. San Raffaele
Via Olgettina 58, I–20132 Milan (Italy)
Tel. +39 02 21560205, Fax +39 02 21560220, E-Mail vescovi.angelo@hsr.it

notions regarding the functional properties of the adult neural stem cells and discuss findings in the expanding area of trans-germ layer conversion, with emphasis on the neural stem cell.

Introduction

In mammals, the vast majority of the bodily cells derive from early totipotent precursors of the inner cell mass of the blastocyst which are called embryonic stem (ES) cells. At around the time of gastrulation, distinct groups of precursors segregate from these ES cells into specific cell lineages found in the three main germ layers, that is ectoderm, mesoderm and endoderm. Gradually these precursor will mature into organ- and tissue-specific somatic stem cells (SCs) [Faust and Magnuson, 1993] which account for the progressive growth of the tissues mass during development. SCs decrease in number with the tissue approaching maturity. In tissue where continuous and extensive cell replacement is clearly documented (the hemopoietic system, the epidermis, the intestinal epithelium) they reach a steady state level that is maintained throughout adult life. This phenomenon essentially relates to the function of maintenance and repair that SCs are thought to carry out in those tissues and organs. Conversely, the mature mammalian CNS has long been perceived as being incapable of any significant cell turnover and it was thought to be devoid of stem cells. This view has progressively changed in the last few decades, and the existence of de novo neurogenesis and the presence of stem cells in the adult mammalian CNS have emerged in recent times. Currently, the forebrain subventricular zone lining the forebrain ventricles – a remnant of the embryonic subventricular zone – and the subgranular zone of the dentate gyrus of hippocampus have been recognized as germinative zones of the adult brain. The SVZ is thought to be the main stem cell compartment in the mature brain. From this area, adult neural stem cells (ANSCs) have been isolated that have been cultured and expanded while retaining unchanged functional attributes for an extended time in vitro [reviewed in Cameron and McKay, 1998; Temple and Alvarez-Buylla, 1999; Kuhn and Svendsen, 1999; Gage, 2000].

Although it has been implied that ANSCs possess a significant degree of functional plasticity, both in terms of growth and expansion capacity and differentiation ability, it has also been generally assumed that the fate of these cells was restricted to the generation of the three major brain cell types, namely neurons, oligodendrocytes and astrocytes. Here, we examine some basic notions regarding the functional properties of ANSCs, particularly focusing on SVZ-derived stem cells, as prefatory to the discussion that concerns the extreme plasticity of these cells, which has recently emerged as one of their more prominent functional characteristics. The capacity of ANSCs to generate cells that are normally derived from embryonic germ layers other than the neuroectodermal one will also be discussed.

The Adult Stem Cell

In a general view, SCs proliferate actively during development and, through a process of progressive lineage restriction, eventually generate terminally differentiated, tissue-specific cells. While diversification into distinct mature cell types is complete at or soon after birth, many tissues in the adult organism physiologically undergo continuous cell turnover and repair and must therefore embody a 'reservoir' of rather pliable SCs. These are often relatively quiescent or slow proliferating cells. However, SCs retain a significant ability to increase their activity to replace dead and/or injured cells, often through the generation of an intermediate, fast-proliferating cell population called the transit amplifying cell compartment [Potten and Loeffler, 1990; Loeffler and Potten, 1997].

The notable effort that has gone into defining the specific molecular and antigenic markers for the various types of adult stem cells notwithstanding, the identification of many SCs is still carried out on an operational basis and relies on the retrospective assessment of critical, defining functional characteristics in candidate cells, with a remarkable exception being represented by hemopoietic stem cells, for which a wide array of markers is already available.

The most widely accepted operational definition of a stem cell [Potten and Loeffler, 1990; Loeffler and Potten, 1997] identifies them as undifferentiated cells (i.e. lacking antigenic markers typical of mature cells) displaying a significant proliferation potential that, in turn, is inextricably linked to their extensive (possibly life-long) self-maintenance capacity. Further, SCs are viewed as multipotent cells – i.e. they can give rise to a wide array of differentiated offspring specific of the tissue in which they reside – and should possess the capacity to regenerate/repair the tissue of origin, even following extensive damage.

Self-maintenance is the capacity of a cell to perpetuate itself and, at least in invertebrates, it is perceived as the property of single cells to undergo rigid asymmetric divisions that result in the production of one cell identical to their mother and of one more differentiated sister cell. Conversely, in vertebrates, self-renewal is viewed as the global property of stem cell populations, in which an equal balance between the number of symmetric divisions that generate two stem cells or, alternatively, two differentiated cells at each cycle is present [Morrison et al., 1997; Fuchs and Segre, 2000]. Shifting this equilibrium in favor of the first or second type of division will either determine an increase or, alternatively, a decrease in the number of stem cells within a given compartment. This mechanism provides a regulatory system by which the size of a stem cell population and the number(s) of their differentiate progeny can be varied in response to changes in the extracellular environment that can follow injury or the onset of pathological conditions.

Multipotency is the ability of a single cell to generate many different types of mature progeny. In principle, multipotent SCs should be able to give rise to all the cell types that constitute their tissue. Thus, different types of stem cells differ in their developmental potential (tissue specificity). In vivo, this criterion may be difficult to assess and stem cells may appear to generate only a subset of the differentiated cell types of a given tissue. However, the potential of SCs may be more thoroughly unraveled by in vitro assays [Reynolds and Weiss, 1992; Gritti et al., 1996; Palmer et al., 1999; Doetsch et al., 1999; Gritti et al., 1999; Mi and Barres 1999], in which candidate cells are given oportunities that may not be available in vivo.

A remarkable characteristic of SCs is that they are located within restricted tissue regions that warrant a cytoarchitectural and biochemical confinement which, in turn, may establish a specific 'stem cell niche'. Within this niche, peculiar conditions can be maintained in order for SCs to retain their peculiar functional attributes and, particularly their life-long ability to self-renew and generate many mature progeny. In some tissues like the liver and the hemopoietic system the anatomical location of the stem cell niche and its relationship to the tissue-specific stem cells are not clearly understood. In other systems, like the epidermis of non haired skin, the small intestine and the hair follicle, the niche is spatially defined and the stem cells homing there can be efficiently recognized on the base of their morphology and position within the niche itself [Fuchs and Segre, 2000; Weissmann, 2000]. Little is known about the niches that are present in the germinative zone of the adult brain. However, the notion of a stem cell niche is supported by both anatomical data [Doetsch et al., 1997, 1999; Palmer et al., 2000] and transplantation experiments (see next section). It is becoming clear that different types of molecules and different cell types contribute to the neural stem cell microenvironment. Some of the molecular signals that are required for the neurogenic lineage to be maintained throughout life are now being unraveled (see next section).

Adult Neural Stem Cells: Their Identity and Plasticity

The cytoarchitectural complexity that is the distinctive hallmark of the mammalian brain parenchyma develops from a much simpler, neuro-epithelial embryonic structure, that is the neural tube. This contains undifferentiated, mitotic neural progenitors that initially reside in the luminal cell layer or ventricular zone (VZ). As development proceeds, a new germinal layer appears beneath the original VZ, which is called the subventricular zone (SVZ). The thickness of the VZ decreases gradually and, eventually, this is reduced to a continuous cell monolayer which lines the ventricular cavities in the adult CNS (ependymal layer). Importantly, a remnant of the primitive SVZ persists in the forebrain throughout adulthood. The adult SVZ is constituted by actively mitotic cells which a large body of evidence now identifies as being the neural precursors responsible for the persistent and intense neurogenetic activity within this area. This has raised questions as to the existence of niches in the mature brain within which a peculiar set of signals supports neurogenesis by 'nurturing' a specific subset of neural stem cells throughout life [reviewed in Gage et al., 1995; Peretto et al., 1999; Temple, 1999]. In fact, bona fide neural stem cells have been isolated from the adult SVZ and hippocampus [Reynolds and Weiss, 1992; Richards et al., 1992; Gage, 1998] and different groups have suggested that a multipotent stem cell compartment resides within the adult, periventricular forebrain region and in the hippocampal dentate gyrus [Gritti et al., 1996; Morshead et al., 1994; Palmer et al., 2000]. The current view identifies the periventricular forebrain area as the principal stem cell pool of the adult mammalian CNS. From this location, large numbers of cells are generated in both the neonatal [Luskin, 1993] and adult [Doetsch et al., 1999] brain that, in turn, migrate along a restricted pathway in order to reach the olfactory bulb (OB) where they eventually differentiate into two different types of mature neurons, that is granular and periglomerular interneurons. These stem cells can be isolated from the SVZ and/or the ependymal layer and, when propagated in long-term cultures in a

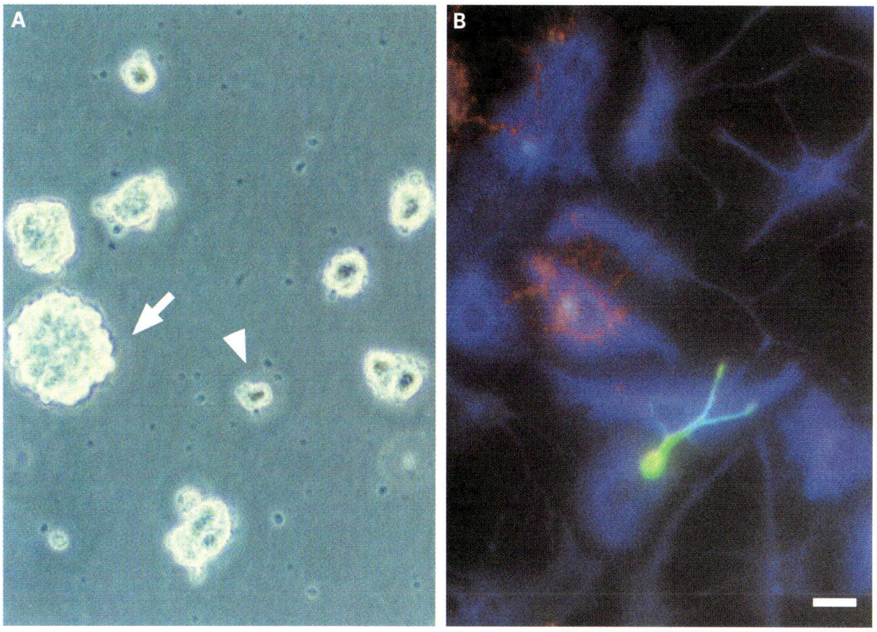

Fig. 1. Multipotent neural stem cells isolated from the adult mouse brain. The figure shows an example of adult neural stem cells that were isolated from the forebrain periventricular region. **A** In low-density cultures, in serum-free medium and in the presence of EGF and FGF2 cells actively proliferate and can be seen at early stages as hypertrophic, phase bright cells (arrowhead) which, eventually, give rise to spherical clusters of cells called neurospheres (arrow). **B** Upon growth factor removal these cells will differentiate into neurons, astrocytes and oligodendrocytes, as shown by specific triple immunolabelling with antibodies raised against β-tubulin (yellow), GFAP (blue) and O4 (red).

serum-free medium, they retain extensive self-renewal, multipotency and stable functional features over time [Reynolds and Weiss, 1992; Richards et al., 1992; Gritti et al., 1996; Morshead et al., 1994; Johansson et al., 1999] (fig. 1). By this system, a steady expansion of the stem cell population is achieved which strictly depends on the pattern of mitogenic signals to which cells are exposed. An intensive expansion of the stem cell number occurs in the simultaneous presence of EGF and FGF2 [Gritti et al., 1999] and indicates that symmetric divisions in which a stem cell gives rise to two daughter stem cells occurs with a high frequency under these conditions [Morrison et al., 1997]. When only one of the two factors is used as mitogen, the same type of ANSCs can revert to a much slower rate of expansion, since, under these conditions, the frequency of symmetric divisions yielding two differentiated cells at each cycle are significantly increased [Gritti et al., 1999]. This discovery underlines a critical aspect of ANSCs' physiology, namely their ability to vary their self-renewal activity in relation to changes in the surrounding environment.

That the fate of ANSCs is under tight environmental control is also demonstrated by the observation that, upon removal of growth factors (GFs) from the culture medium, cells undergo growth arrest and differentiate in a mix of neuronal and glial cells. Strikingly, the timing by which these mature CNS lineages are produced is reminiscent of the situation observed in vivo within most embryonic neurogenetic regions, so that neurons are generated first, followed by astroglial cells and oligodendrocytes.

Further, the fate of the various differentiated stem cell progeny clearly depends on extracellular signaling. Thus, FGF2 prevents the differentiation/maturation of glial and neuronal/glial precursors in serum-free medium, while the addition of low concentration of serum is a necessary step to improve neuronal maturation and the expression of glial antigens [Vescovi et al., 1993]. Also, neurotrophins play a fundamental role in regulating both the survival and differentiation of stem-cell-generated neurons in culture [Vicario-Abejon et al., 1995; Bibel and Barde, 2000], while several other molecules (BMPs, PDGF, FGFs, LIF, CNTF) control the proliferation, the fate selection and the differentiation of ANSC-derived astro- and oligodendroglial precursors [Barres et al., 1994; Raff et al., 1988; Mabie et al., 1997; Gross et al., 1996; Mi and Barres, 1999].

An intriguing aspect of ANSCs plasticity is the ability to modify their fate in response to instructive extracellular cues. For instance, it has been shown that the ratio of neuronal versus glial differentiation in stem cells can be biased by various combinations of epigenetic signals. Thus, the final outcome of the stem cell differentiation process in murine ANSCs can be directed by PDGF toward the acquisition of a neuronal fate in as many as half of their differentiated progeny [Johe et al., 1996], a phenomenon similar to that observed in fetal human NSCs cultured in the presence of LIF [Galli et al., 2000b]. On the contrary, ciliary neurotrophic factor (CNTF) appears to act instructively on murine ANSCs to generate a 'larger than normal' proportion of glial cells [Johe et al., 1996].

Concerning the neurotransmitter phenotype that the stem cell-derived neuronal progeny can acquire, it is known that the large majority of these neurons display a gabaergic or glutamatergic phenotype when differentiation is triggered by the removal of GFs in serum-free conditions, in vitro. However, there is clear evidence that a significant degree of plasticity exists in ANSCs with respect to their choice of the neurotransmitter phenotype. Phenotypes of therapeutic relevance, such as the catecholaminergic one, can be induced by exposing the ANSC progeny to peculiar differentiation conditions that entail the use of medium conditioned by astrocytes as well as FGF2 and other factors like IL-1 [Daadi and Weiss, 1999; Yan et al., 2001; Wagner et al., 1999].

Environmental modulation of ANSCs' activity can also take place in vivo. Thus, a dramatic increase in newborn cells is observed in the adult brain following intraventricular administration of EGF, FGF-2 [Wagner et al., 1999; Craig et al., 1996], brain-derived neurotrophic factor (BDNF) [Zigova et al., 1998; Pencea et al., 2001a, b] or periferal infusion of insulin growth factor-I [Aberg et al., 2000]. Interestingly, while FGF2, BDNF and IGF-I induce an increase in the number of neurons generated, EGF seems to bias differentiation toward the glial lineage [Kuhn et al., 1997] and bears an inhibitory effect on the progression of SVZ-derived cells through their usual migratory routes, inducing their displacement from the SVZ into the adjacent striatal region [Wagner et al., 1999]. This indicates that different cell behaviors in terms of survival, proliferation, fate choice and cell migration, can be induced by different combinations of extracellular signals, in a rather complex manner, in vivo. A recent confirmation of these phenomena has emerged from the work of Fallon et al. [2000], in which the infusion of TGFα in the striatum of 'parkinsonian' rats triggered the proliferation of SVZ cells as well as the concomitant migration of their progeny toward the site of injection. Furthermore, this was followed by the differentiation of some of the newly born cells into TH-immunoreactive neurons.

The findings discussed above lead to the consideration that epigenetic signals influence critical neurogenetic steps during which growth, fate determination, differentiation and maturation are finely regulated. Therefore, the rate and extent by which cells found in the mammalian CNS are generated is subjected to a flexible regulation that can be viewed as a clear example of neural plasticity thanks to which the production of varying quantities and type(s) of mature brain cells can be attained.

A plethora of recent studies are now beginning to unravel the functional characteristic and nature of ANSCs in vivo. The cellular composition of the periventricular region has now been thoroughly analyzed and at least four cell types have been characterized: ependymal cells, type A cells (neuroblasts), type B cells (astrocytes) and type C cells (immature precursors) [Doetsch et al., 1997]. Separate studies have shown that the ependymal cells that line the luminal surface of the adult ventricular wall [Johansson et al., 1999] and the astrocytes that resides in the adjacent SVZ (type B cells) [Doetsch et al., 1999] may function as the source of the multipotent ANSCs observed in vitro. To provide direct evidence that type B cells do give rise to neurons, Doetsch et al. [1999] used a transgenic mice expressing the avian leukosis viral (ALV) receptor driven by the promoter/enhancer of the glial fibrillary acidic protein (GFAP) gene. When these mice were infected with the ALV encoding alkaline phosphatase (AP), AP-expressing SVZ astrocytes were generated that produced a rapidly dividing type C cells which, in turn, gave rise to neuronal type A cells that migrated to the OB. By using a similar experimental approach, Seri et al. [2001] recently suggested that the granule neurons of the hippocampus, the other major neurogenetic region of the adult brain, are generated by astrocytic-like stem cells through a transit amplifying cell population (type D cells).

Of note, it has recently been shown that adult SVZ-derived astrocytes grown as monolayer do resume neural stem cells features when re-cultured in presence of both FGF2 and EGF [Laywell et al., 2000]. Furthermore, SVZ astrocytes express radial glial markers [Hartfuss et al., 2001; Noctor et al., 2001] indicating they may retain an immature 'radial glia-like' phenotype that could be functionally related to the type B astrocytes of the SVZ. These results show that multipotent stem cells of the adult brain are very likely represented by a small subset of astroglial-like SVZ cells. However, the possibility that ependymal cells act as stem cells, directly or through the generation of SVZ cells, remains a possibility that needs further clarification.

Neurogenesis in the Adult CNS: The Importance of the Brain Microenvironment

The molecular signals that are required to warrant the maintenance of a stem cell pool endowed with permanent neuronogenetic capacity in the adult brain are poorly understood. Yet, one may borrow the notion of 'stem cell niche' that was developed in other areas of investigations on somatic stem cells and apply it to the neural tissue. This leads to speculate on the existence of specific domains within the adult brain, an idea that is supported by transplantation experiments. For instance, adult rat hippocampal progenitor cells grafted into neurogenic (the SVZ) or non-neurogenic (the cerebellum) regions of adult hosts specifically give rise to neurons only in the former [Suhonen et al., 1996]. Further, SVZ cells transplanted in the SVZ of a recipient animal generate large numbers of new neurons [Lois and Alvarez-Buylla, 1994; Doetsch et al., 1999] while the very same cells grafted to non-neurogenetic CNS regions (cortex and striatum) give rise mainly to astroglia [Herrera et al., 1999]. Likewise, homotopic transplantation of spinal cord-derived ANSCs expanded in vitro [Shihabuddin et al., 1997] into adult rats resulted in the production of glial cells only. Conversely, after heterotopic transplantation into the dentate gyrus of the hippocampus, a known neurogenetic region in the adult brain [Gage, 1998], the same cells integrated in the granular layer and differentiated into neurons, whereas engraftment into other hippocampal regions resulted, once more, in the sole production of glial cells [Shihabuddin et al., 2000]. These data clearly show that extracellular cues may direct the acquisition of specific differentiated fates in neural precursors in vivo, and that only specific brain areas remain permissive for continued neurogenesis throughout adult life.

Many efforts have gone into unraveling the molecular signals that drive neurogenesis and in trying to determine their underlying mechanism of action. The scenario that has emerged so far indicates that molecules of various nature (secreted factors, membrane proteins, extracellular matrix components) as well as different cell types, act and interact to contribute to the specification of particular domains in the neural stem cell's microenvironment. Studies on cultured SVZ cells point to EGF and FGF2 as essential components of a potential 'neural stem cell niche'. As discussed above, these growth factors can support the proliferation and self-renewal of ANSCs in vitro, while the infusion of EGF or FGF2 into the forebrain ventricles causes the expansion of the SVZ cell population in vivo [Craig et al., 1996; Kuhn et al., 1996]. The importance of signaling through the EGF-receptor family in adult neurogenesis in vivo is further supported by the observed reduced dorsolateral SVZ cell proliferation found in the TGFα-knockout mice [Tropepe et al., 1997]. Other factors have also been recently implicated as potential regulators of stem cell activity in the SVZ. Among them are ephrins and their cognate tyrosine kinase receptors. In fact, perturbation of the ephrins' signaling significantly disrupts both the pattern of migration of SVZ stem cell-derived type A progeny and the proliferation of the SVZ cells themselves [Conover et al., 2000]. On the other hand, noggin, a polypeptide that binds bone morphogenetic proteins (BMPs) and prevents the activation of the BMP receptors [Wilson and Hemmati-Brivanlou, 1997] appears to be involved in regulating SVZ cells activity. Lim et al. [2000] recently proposed that noggin, as produced by ependymal cells, may antagonize BMP autocrine signaling in type B cells. Since the latter normally blocks the neurogenic pathway, this may create a neurogenetic environment in the adjacent SVZ and drives SVZ cells toward the acquisition of a neuronal identity.

The permissive environment provided by extracellular matrix, cell surface molecules and special supporting cells (i.e. radial glia) allows the displacement of neuronal precursors from the site of genesis to the site of their full differentiation during development. This is somehow reproduced within those regions of the adult brain that are endowed with structural plasticity and/or the capacity for life-long active neurogenesis. For example, the polysialylated, 'embryonic' isoform of the neural cell adhesion molecule N-CAM (PSA-NCAM), which plays an important role in cell migration and cell shaping [Bonfanti et al., 1992; Chazal et al., 2000; Hu et al., 1996], is selectively expressed on the membrane of newly-generated cells in the hippocampus [Seki and Arai, 1993] and in the SVZ [Bonfanti and Theodosis, 1994]. The expression of PSA-NCAM is crucial for SVZ-generated neuroblasts, that migrate tangentially inside tube-like structures formed by specialized glial cells to join a restricted pathway (the rostral migratory stream) by means of homotypic interaction without radial glia or axonal guides (chain migration) [Lois and Alvarez-Buylla, 1996]. The formation of a specific microenvironment in the stem cell niche may also entail the presence of specific cell types that provide mechanical and/or trophic sustenance to ANSCs. In a way somehow similar to that observed with stromal and hemopoietic stem cells in the bone marrow, radial glial cells provide support for neuronal migration and supply instructive and neurotrophic signals that are required for survival, proliferation and differentiation in the developing CNS [reviewed in Chanas-Sacre et al., 2000]. Likewise, in the adult brain, short radial glia-like cells are known to be present in the hippocampal dentate gyrus [Cameron et al., 1993], while SVZ astrocytes (type B cells) that surround the cells migrating towards the OB express embryonic-like cytoskeleton proteins such as vimentin and nestin [Jankovski and Sotelo, 1996; Peretto et al., 1999; Lim and Alvarez-Buylla, 1999].

Once the appropriate set(s) of epigenetic cues is established/maintained within the niche, self-renewal of the neural stem cell compartment ultimately depends on the integration of these signals, that is performed by cell intrinsic regulators. As recently suggested, NSCs may behave like dynamic microprocessors in which different inputs are integrated as a function of time, in a way affecting the probability of different outputs [Anderson, 2001]. The identity of the intrinsic mechanisms and of the signaling cascades that become activated upon interaction with epigenetic factors is complex and still poorly understood. Nuclear factors controlling gene expression in stem and progenitor cells, molecules involved in the control of asymmetric divisions of stem cells and clock mechanisms that set the number of division rounds within the population are among the most likely candidate as cell intrinsic regulators [reviewed in Watt and Hogan, 2000]. In this view, it is intriguing to note that some of the transcription factors and of their downstream effectors that are involved in the regulation of cell cycle and proliferation in the stem cell niches of the epidermis and intestinal epithelium [reviewed in Watt and Hogan, 2000] have also been found in the developing and adult CNS, as well as in neural-derived primary cultures and cell lines [Hughson et al., 1998; Cho and Dressler, 1998; Galceran et al., 2000; Satho and Kuroda, 2000]. This suggests that an evolutionarily conserved set of cell autonomous mechanisms may be acting in different types of somatic stem cells by fulfilling the ultimate function of regulating the size of the stem cell compartment and the varying output of terminally differentiated cells that are contributed to adult tissues throughout adult life.

Trans-Differentiation (Metaplasia) of Somatic Stem Cells

SCs have been isolated and characterized from many adult mammalian tissues, including the bone marrow, muscle, skin, gut, and CNS. These cells are responsible for the cellular turnover elicited by physiological cell depletion, pathological situations or injury. As a consequence, the differentiation potential of SCs has been thought to be restricted to the production of mature cells that belong to the tissue in which the SCs reside. Several findings are now challenging this dogmatic notion.

In 1998, Ferrari et al. showed that bone marrow precursors can differentiate into skeletal muscle cells following chemically-induced damage of the adult tibialis anterioris in mice. These finding documented the intra-germ layer conversion capacity of adult stem cells, that is the production of mature progeny normally found in a tissue that is different from that in which the stem cell resides but that derives from the same embryonic germ layer. A similar study reporting the ability of muscle-derived cells to display hematopoietic features [Jackson et al., 1999] has now ben revisited, since fractionation of muscle cells on the basis of hematopoietic markers has demonstrated that the muscle-derived population bearing the potential to repopulate all major peripheral blood lineages exclusively express the hematopoietic stem cell antigens Sca-1 and CD45, thus being hematopietic (and not myogenic) in nature [McKinney-Freeman et al., 2002]. In 1999, Bjornson et al. demonstrated that clonally-derived ANSCs (ectodermal derivatives) could give rise to hematopoietic cells (mesodermal derivatives) when injected into sublethally irradiated adult mice [Bjornson et al., 1999]. This provided the first demonstration that somatic stem cells that were derived from a given germ layer could generate cells of a distinct embryonic origin (trans-germal conversion). Later on, this striking developmental flexibility began to emerge as a more general feature of other types of adult somatic stem cells. Thus, bone marrow precursors were proved to contribute to the regeneration of organs of endodermal origin such as the liver, and to be capable of converting into neural cell lineages [Petersen et al., 1999; Theise et al., 2000; Krause et al., 2001; Mezey et al., 2000; Brazelton et al., 2000]. Similarly, mesodermal derivatives such as mesenchymal stem cells can give rise to cells bearing morphological and antigenic properties of astrocytes and neurons, both in vivo [Kopen et al., 1999] and in vitro [Sanchez-Ramos et al., 2000; Woodbury et al., 2000].

The unexpected plasticity of adult CNS stem cells was quite astounding, particularly considering that the nervous system had always been depicted as the most 'quiescent' amongst adult tissues, at least in terms of cell turnover in adult life. Yet, such extraordinary capability of ANSCs was soon confirmed and extended by the demonstration that, upon injection into the mouse blastocyst, ANSCs could integrate into many different tissues derived from the three main germ layers [Clark et al., 2000]. However, in this work the contribution of ANSCs to two

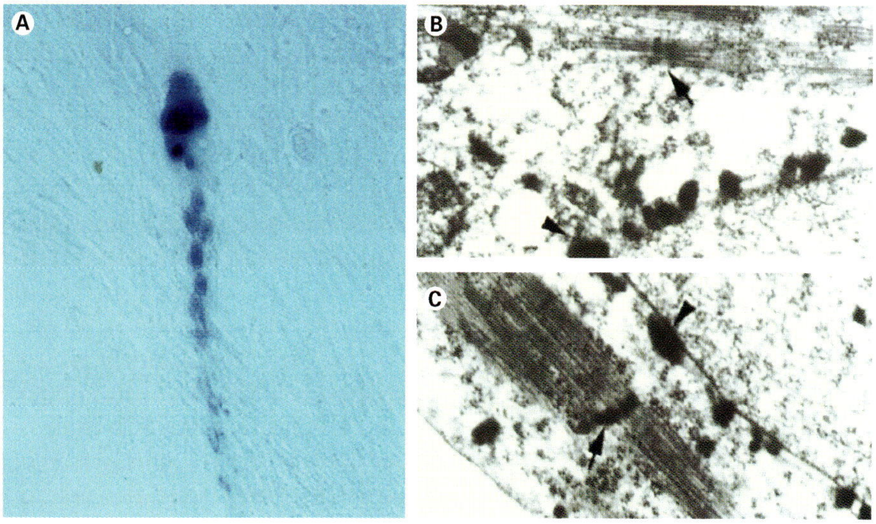

Fig. 2. Multipotent neural stem cells from the adult mouse brain give rise to skeletal muscle cells. ANSCs were isolated and cloned from transgenic MLC3F mice in which the nuclear expression gene lacZ, encoding for the bacterial enzyme β-galactosidase, is driven by the promoter of the mammalian heavy chain myosin gene (HCM). ANSCs do not normally express HCM or any other muscle marker [Galli et al., 2000a]. When they are cocultured with myogenic cells they fuse into polynucleated fibers and express HCM. This is clearly shown in panel (**A**), in which the blue staining upon reaction with X-gal indicates that this multinucleated fibre derives from neural cells that have switched on the HCM gene. **B**, **C** Electron microscopy analysis of ANSCs induced to generate skeletal muscle cells reveals the presence of sarcomeric structures in cells of neural origin (as demonstrated by the perinuclear deposition of the electron-opaque X-gal reaction product) (arrowheads). The presence of M (**B**, arrow) and Z bands (**C**, arrow) in these cells warrants the generation of mature muscle cells from adult brain stem cells.

of the major mesodermal lineages such as blood and the skeletal muscle was not observed. Our group demonstrated soon thereafter that ANSCs or human fetal neural stem cells do possess the ability to differentiate into skeletal muscle, both in vivo and in vitro [Galli et al., 2000a] (fig. 2). The apparent discrepancy between our work and that of Clarke and colleagues is likely justified by the different cell systems and assays used to investigate this phenomenon. In the former, ependymal-derived ANSCs were injected into an embryonic environment whereas, in our experiments, SVZ-derived adult stem cells underwent trans-germ layer conversion in the context of an adult, damaged tissue. Thus, while in a regenerating adult tissue the 'inducing' cues will probably direct trans-differentiation toward the generation of local, tissue specific lineages [Galli et al., 2000a], distinct and antagonistic signals are very likely to act upon exogenous cells as injected into a blastocyst. These signals will compete to direct the fate of the transplanted cells toward many alternative pathways that are developing, simultaneously, in the embryonic environment. In the latter situation, the final fate acquired by the exogenous cells will then emerge as the result of their exposure to numerous, yet unpredictable factors. To compound the problem, the overall composition pattern of these factors will depend upon the precise stage of blastula development and on a series of unpredictable positional cues that relate to the initial site of integration of the transplanted cells. This situation is likely to cause a significant degree of variability in the outcome of the blastula experiments which will require the analysis of a significant number of grafts to allow for categorical conclusion to be reached as to the actual overall potential of the donor cells. Recent data from Pipia et al. [2000] seem to confirm this view since, when these authors injected neural stem cells into blastocysts, the very first lineage to be colonized by the neural cells was, indeed, the hemopoietic one.

The many reports describing CNS stem cell plasticity underline the need for rather peculiar conditions in order for SCs to express what appears to be a latent, generalized developmental potential(s). Two main conditions have been identified as necessary for CNS stem cell conversion to occur both in vivo and in vitro. First, the stem cells ought to be in a highly undifferentiated state. In fact, we reported that the mitotic state of ANSCs does represent an important element that can influence the rate by which

ANSCs can convert into non-CNS, myogenic cells. Thus, ANSCs need to be in a state of active proliferation in order to undergo efficient conversion into non-neural cells either in vivo or in vitro [Bjornson et al., 1999; Clark et al., 2000; Galli et al., 2000a]. As described by Galli et al. [2000a], when ANSC-derived terminally differentiated progeny (neurons and glia) were exposed to cues that induce the myogenic phenotype, almost no conversion to skeletal muscle was observed.

Second, it can be speculated that, to undergo trans-germ layer conversion ANSCs ought to be exposed to microenvironments that are rich in instructive cues. CNS stem cells have been shown to repopulate both the early developing gastrula [Clark et al., 2000] as well as the regenerating hemopoietic system [Bjornson et al., 1999] and muscle [Galli et al., 2000a], while bone marrow cells regenerate the damaged muscle [Ferrari et al., 1998], liver [Petersen et al., 1999], gastrointestinal tract, lung or skin [Krause et al., 2001]. These observations suggest that the host tissue limits the number of donor cells that can contribute to it in physiological condition, and that instructive cues may only become available during the regenerative phase that follows an injury in a target adult tissues or during embryonic development. Expression of CD34 is increased in mice shortly after transplantation in the marrow of lethally irradiated recipients of single bone marrow cells, consistent with this molecule being involved in homing [Krause et al., 2001]. This observations are consistent with the low frequency of phenotypic conversion that is reported in most of the published studies and that is one of the critical parameter to be considered in assessing transdifferentiation [Anderson et al., 2001].

The fact that only ANSCs but not their differentiated neuronal or glial progeny are susceptible of giving rise to non-neural cells [Galli et al., 2000a] highlights an important difference between this phenomenon and transdifferentiation as conceived in its classical form. In fact, transdifferentiation implies the ability of differentiated cells to acquire a new identity by turning off one set of lineage-specific genes and activating genes of another differentiated cell type. Conversely, trans-germ layer differentiation seems to reflect the de novo expression of a broader developmental potential of ANSCs that may become evident only under peculiar conditions. Thus, from a more formal perspective the findings so far discussed fit more appropriately into a class of biological phenomena called metaplasia [Slack, 1986].

It is worth noting that no molecular and biochemical overlapping of specific neural and non-neural markers was ever observed in ANSCs and their non-neural progeny. Thus, markers of hematopoietic precursors were not detected in ANSCs before they were injected into injured animals [Bjornson et al., 1999] and expression of muscle determination factors or other muscle differentiation markers such as MyoD, Myf5, myogenin and myosin was never observed in ANSCs prior to their exposure to 'myogenic' environments [Galli et al., 2000a] (fig. 2). Finally, lack of expression of mesodermal and endodermal antigens, such as for example TROMA-1, was detectable in ANSCs before injection into the developing blastula [Clark et al., 2000], yet these cells can give rise to derivatives from both these germ layers. In all these examples, ANSCs expressed appropriate neural stem cell antigens such as nestin in their undifferentiated state and neuronal and glial markers following terminal differentiation.

An important issue arises from the observed, unexpected plasticity of SCs, which concerns the mechanisms that underlie to this striking phenomenon and to the identification of the signals found in the 'inductive' environments. Unfortunately, very little is known as to the identity of the factors responsible for this peculiar transition in SCs' fate. Nevertheless, based on our recent work showing the generation of skeletal muscle from ANSCs it is possible to infer some speculative hypothesis. By these experiments, we have established the first in vitro model in which neuro-mesodermic conversion can be elicited and studied quantitatively, under controlled conditions. We used ANSCs isolated from the forebrain periventricular region and cultured at low density, in serum-free-, mitogen-containing culture medium. Under these culture conditions these cells are functionally stable over time in culture [Gritti et al., 1999]. The possibility that in vitro manipulation and long-term culturing may cause cell reprogramming is real [Kondo and Raff, 2000; Palmer et al., 1999]. However, the transdifferentiation potential shown by ANSCs was not a property acquired in vitro [Anderson et al., 2001], since acutely dissociated periventricular cells underwent myogenic conversion when co-cultured with myogenic cells [Galli et al., 2001a]. Moreover, although donor ANSCs were not prospectively isolated [Uchida et al., 2000], the use of clonally-derived ANSCs ruled out a possible contamination of our donor cell population with stem cells/progenitors of different origin. Using this system, it has emerged that this conversion could be observed only when ANSCs were co-cultured with C2C12 myogenic cells or with primary myoblasts, and never with non-myogenic cells. Interestingly, conversion required direct cell-cell contact and did not take place when a porous membrane physically separated neural and myogenic cells, nor was it observed when

ANSCs were exposed to myogenic cell-derived extra-cellular matrix or to medium in which muscle cells had been grown. Furthermore, when ANSC underwent myogenic induction as undissociated neurospheres (clusters of cells formed by the clonal-derived progeny of an individual stem cell), the proportion of cells that underwent conversion to skeletal muscle dropped by almost 80% as compared to the same number of ANSCs that were dissociated before induction. This took place despite the fact that ANSCs neurospheres rapidly spread onto myoblasts – so that the cells in the sphere were in contact with the inducing C2C12 cells. The occasional myogenic conversion observed in 'clustered' co-cultures occurred exclusively at the exterior of the neurosphere and never within its core, which, accordingly, contained solely neural cells. This was somehow expected, since a neurosphere is a highly heterogeneous population of cells, containing stem cells as well as more committed progenitors and even terminally differentiated cells, these last usually homing in the core of the sphere. Thus, when neural cells are clustered together, neural-to-neural signals override the myoblast-derived myogenic signals, blocking neuro-myogenic conversion. Therefore, in analogy to what observed during the dedifferentiation of retinal pigmented epithelial into lentoids [Kodama and Eguchi, 1994], loss of cell-to-cell communication amongst ANSCs emerges as one of the major determinant influencing the onset of neural-to-muscle conversion.

Based on these findings, we can infer that, at least in the specific case of neuroectodermal-mesodermal conversion, cell-to-cell contact between the inducing (muscle) and the induced (neural) cells must occur. Since cell fusion did not appear as a prerequisite for neuro-myogenic conversion in our system, the need for direct cell-to-cell interaction underlines the necessity of a direct exchange of information between the 'inducing' and the 'induced' cells. This may imply the interaction of cell surface receptors and ligands, the formation of gap junction structures, as well as the involvement of short-range acting molecules. However, we must emphasise that the involvement of secreted factors in neuro-myogenic conversion cannot be ruled out, although it can safely be argued that diffusible molecules alone cannot enforce such an extreme change in cell fate. This is further confirmed by the observation that culturing ANSCs in the sole presence of hematopoietic growth factors and cytokines fails in triggering conversion of neural cells into blood [Bjornson et al., 1999]. Finally, it is evident that a fine interplay between antagonistic cues occurs during neuro-myogenic conversion. In fact, the induction of the muscle fate elicited upon ANSCs is counteracted by some type of 'neutralising' signalling occurring between neural cells that are in direct contact, a phenomenon that may be interpreted as a classical 'community effect'. Thence, a cohort of instructive signals rather than a single effector may control the shift from a brain-specific fate to a mesodermal one. In vivo, these signals can be found either in the extracellular microenvironment that has been perturbed by an injury or may be elicited through a direct cell-to-cell interaction between the host and the donor cells.

Along with the influence of the external milieu in controlling stem cell flexibility, cell-autonomous modifications may be involved in the trans-differentiation (metaplastic) process. Initial evidence is now becoming available regarding this specific aspect. For instance, the gene called Pax7 has been shown to control the differentiation of adult muscle satellite cells into hemopoietic cells, since Pax7 deletion increased the capacity of adult muscle stem cells to undergo hematopoietic differentiation [Seale et al., 2000]. Also, PTF-1 and PDX-1 expression in pancreatic acinar cells has been associated to the neo-generation of endocrine islet cells, and exocrine cells can trans-differentiate and acquire characteristics typical of precursors active during beta-cell neogenesis [Rooman et al., 2000]. Finally, similarities between the gene expression patterns observed during embryonic development and during transdifferentiation have been documented for the cornea-lens transdetermination phenomenon [Kodama and Eguchi, 1994]. This implies that the different processes of embryogenesis, regeneration and transdifferentiation are likely highly inter-related at the molecular level.

Conclusions

As of recent, the fact that many SCs possess the capacity to reactivate an apparently dormant set of developmental programs under peculiar environmental conditions has emerged quite clearly. The reactivation of these programs does not require genetic manipulation or nuclear transplantation and is driven by a cohort of extracellular cues that may be found both in the developing organism and within many regenerating adult tissues. This reinforces the concept that the cellular and molecular mechanisms responsible for the acquisition of specific cell fate necessary for the appropriate embryonic development of the mammalian body may be reactivated in adult tissue regeneration. The discovery of the trans-germal differentiation phenomenon is fundamental for the development of novel therapeutic approaches and may allow the development of autologous transplantation by which cells from a healthy tissue can be 'trans-engineered' and used for the therapy of an ill part of the body within the same individual. These therapeutic perspectives impose the need of high standards to prove functional transdifferentiation [reviewed by Anderson et al., 2001]. The possibilty to develop such techniques depends on our understanding of the basic mechanisms that underlie to the conversion of cells derived from different germ layers into another. This emphasizes the importance and the impact that studies on mammalian development and on basic stem cell physiology bear for the establishment of novel therapeutic strategies for human diseases that, to date, remain untreatable by conventional therapy.

References

Aberg, M.A., N.D. Aberg, H. Hedbacker, J. Oscarsson, P.S. Eriksson (2000) Peripheral infusion of IGF-I selectively induces neurogenesis in the adult rat hippocampus. J Neurosci *20(8):* 2896–2903.

Anderson, D.J. (2001) Stem cells and pattern formation in the nervous system: The possible versus the actual. Neuron *30:* 19–35.

Anderson, D.J., F.H. Gage, I.L. Weissman (2001) Can stem cells cross lineage boundaries? Nat Med *7:* 393–395.

Barres, B.A., M.C. Raff, F. Gaese, I. Bartke, G. Dechant, Y.A. Barde (1994) A crucial role for neurotrophin-3 in oligodendrocyte development. Nature *367:* 371–375.

Bibel, M., Y.A. Barde (2000) Neurotrophins: Key regulators of cell fate and cell shape in the vertebrate nervous system. Genes Dev *14:* 2919–2937.

Bjornson, C.R.R., R.L. Rietze, B.A. Reynolds, M.C. Magli, A.L. Vescovi (1999) Turning brain into blood: A hematopoietic fate adopted by adult neural stem cells in vivo. Science *283:* 534–537.

Bonfanti, L. and D. T Theodosis (1994) Expression of polysialylated neural cell adhesion molecule by proliferating cells in the subependymal layer of the adult rat, in its rostral extension and in the olfactory bulb. Neuroscience *62:* 291–305.

Bonfanti, L., S. Olive, D.A. Poulain, D.T. Theodosis (1992) Mapping of the distribution of polysialylated neural cell adhesion molecule throughout the central nervous system of the adult rat: an immunohistochemical study. Neuroscience *49*: 419–436.

Brazelton, T.R., F.M. Rossi, G.I. Keshet, H.M. Blau (2000) From marrow to brain: Expression of neuronal phenotypes in adult mice. Science *290*: 1775–1779.

Cameron, H.A., R. McKay (1998) Stem cells and neurogenesis in the adult brain. Curr Opin Neurobiol *8*: 677–680.

Cameron, H.A., C.S. Woolley, B.S. McEwen, E. Gould (1993) Differentiation of newly born neurons and glia in the dentate gyrus of the adult rat. Neuroscience *56:* 337–344.

Chanas-Sacre, G., B. Rogister, G. Moonen, P. Leprince (2000) Radial glia phenotype: origin, regulation, and transdifferentiation. J Neurosci Res *61:* 357–363.

Chazal, G., P. Durbec, A. Jankovsky, G. Rougon, H. Cremer (2000) Consequences of neural cell adhesion molecule deficiency on cell migration in the rostral migratory stream of the mouse. J Neurosci *20:* 1446–1457.

Cho, E.A.,G.R. Dressler (1998) TCF-4 binds betacatenin and is expressed in distinct regions of the embryonic brain and limbs. Mech Dev *77:* 9–18.

Clark, D.L., C.B Johansson., J. Wilbertz, B. Veress, E. Nilsson, H. Karlstrom, U. Lendahl, J. Frisen (2000) Generalized potential of adult neural stem cells. Science *288:* 1660–1663.

Conover, J.C., F. Doetsch, J.M. Garcia-Verdugo, N.W. Gale, G.D. Yancopoulos, A. Alvarez-Buylla (2000) Disruption of Eph/ephrin signaling affects migration and proliferation in the adult subventricular zone. Nat Neurosci *3:* 1091–1097.

Craig, C.G., V. Tropepe, C.M. Morshead, B.A. Reynolds, S. Weiss, D. van der Kooy (1996) In vivo growth factor expansion of endogenous subependymal neural precursor cell populations in the adult mouse brain. J Neurosci *16:* 2649–2658.

Daadi, M.M, S. Weiss (1999) Generation of tyrosine hydroxylase-producing neurons from precursors of the embryonic and adult forebrain. J Neurosci *19:* 4484–4497.

Doetsch, F., J.M. Garcia-Verdugo, A. Alvarez-Buylla (1997) Cellular composition and three-dimensional organization of the subventricular germinal zone in the adult mammalian brain. J Neurosci *17(13):* 5046–5061.

Doetsch, F., I. Caille, D.A. Lim, J.M. Garcia-Verdugo, A. Alvarez-Buylla (1999) Subventricular zone astrocytes are neural stem cells in the adult mammalian brain. Cell *97:* 703–716.

Fallon, J., S. Reid, R. Kinyamu, I. Opole, R. Opole, J. Baratta, M. Korc,T.L. Endo, A. Duong, G. Nguyen, M. Karkehabadhi, D. Twardzik, S. Loughlin (2000) In vivo induction of massive proliferation, directed migration, and differentiation of neural cells in the adult mammalian brain. Proc Natl Acad Sci USA *97(26):* 14686–14691.

Faust, C., T. Magnuson (1993) Genetic control of gastrulation in the mouse. Curr Opin Genet Dev *3:* 491–498.

Ferrari, G., G. Cusella-De Angelis, M. Coletta, E. Paolucci, A. Stornaiuolo, G. Cossu, F. Mavilio (1998) Muscle regeneration by bone marrow-derived myogenic progenitors. Science *279:* 1528–1530.

Fuchs, E., J.A. Segre (2000) Stem cells: A new lease on life. Cell *100:* 143–155.

Gage, F.H. (1998) Neurogenesis in the adult human hippocampus. Nat Med *4:* 1313–1317.

Gage, F.H. (2000) Mammalian neural stem cells Science *287:* 1433–1438.

Gage F.H., J. Ray, L.J. Fisher (1995) Isolation, characterization, and use of stem cells from the CNS. Annu Rev Neurosci *18:* 159–192.

Galceran, J., E.M. Miyashita-Lin, E. Devaney, J.L. Rubenstein, R. Grosschedl (2000) Hippocampus development and generation of dentate gyrus granule cells is regulated by LEF1. Development *127:* 469–482.

Galli, R., U. Borello, A. Gritti, M.G. Minasi, C. Bjornson, M. Coletta, M. Mora, M.G.C. De Angelis, R. Fiocco, G. Cossu, A.L. Vescovi (2000a) Skeletal myogenic potential of human and mouse neural stem cells. Nat Neurosci *3:* 986–991.

Galli, R., S.F. Pagano, A. Gritti, A.L. Vescovi (2000b) Regulation of neuronal differentiation in human CNS stem cell progeny by leukemia inhibitory factor. Dev Neurosci *22:* 86–95.

Gritti A., E.A. Parati., L. Cova., P. Frolichsthal, R. Galli, E. Wanke, L. FaravelII, D.J. Morassutti., F. Roisen, D.D. Nickel, A.L. Vescovi (1996) Multipotential stem cells from the adult mouse brain proliferate and self renew in response to basic fibroblast growth factor. J Neurosci *16:* 1091–1100.

Gritti, A., P. Frolichsthal, R. Galli, E.A. Parati, L. Cova, S.F. Pagano, C.R. Bjornson, A.L. Vescovi (1999) Epidermal and fibroblast growth factors behave as mitogenic regulators for a single multipotent stem cell-like population from the subventricular region of the adult mouse forebrain. J Neurosci *19:* 3287–3297.

Gross, R.E., M.F. Mehler, P.C Mabie, Z. Zan, L. Santschi, J.A. Kessler (1996) Bone morphogenetic proteins promote astroglial lineage commitment by mammalian subventricular zone progenitor cells. Neuron *17:* 595–606.

Hartfuss, E, R. Galli, N. Heins, M. Gotz (2001) Characterization of CNS precursor subtypes and radial glia. Dev Biol *229:* 15–30.

Herrera, D.G., J.M. Garcia-Verdugo, A. Alvarez-Buylla (1999) Adult-derived neural precursors transplanted into multiple regions of the adult brain. Ann Neurol *46:* 867–877.

Hu, H., H. Tomasiewicz, T. Magnuson, U. Rutishauser (1996) The role of polysialic acid in migration of olfactory bulb interneuron precursors in the subventricular zone. Neuron *16:* 735–743.

Hughson, E., S. Dowler, K. Geall, G. Johnson, M. Rumsby (1998) Rat oligodendrocyte O-2A precursor cells and the CG-4 oligodendrocyte precursor cell line express cadherins, beta-catenin and the neural cell adhesion molecule, NCAM. Neurosci. Lett *251:* 157–160.

Jackson, K.A., Mi, T., Goodell, M.A. (1999) Hematopoietic potential of stem cells isolated from murine skeletal muscle. Proc Natl Acad Sci USA *96(25):* 14482–14486.

Jankovski, A., C. Sotelo (1996) Subventricular zone-olfactory bulb migratory pathway in the adult mouse: cellular composition and specificity as determined by heterochronic and heterotopic transplantation. J Comp Neurol *371:* 376–396.

Johansson, C.B., S. Momma, D.L Clarke, M. Risling, U. Lendahl, J. Frisén (1999) Identification of a neural stem cell in the adult mammalian central nervous system. Cell *96:* 25–34.

Johe, K.K., T.G. Hazel, Muller, M.M. Dugich-Djordjevic, R.D. McKay (1996) Single factors direct the differentiation of stem cells from the fetal and adult central nervous system. Genes Dev *10:* 3129–3140.

Kodama, R., G. Eguchi (1994) Gene regulation and differentiation in vertebrate ocular tissues. Curr Opin Genet Dev *4:* 703–708.

Kondo, T., M. Raff, (2000) Oligodendrocyte precursor cells reprogrammed to become multipotential stem cells. Science *289:* 1754–1757.

Kopen, G.C., D.J. Prockop, D.G. Phinney (1999) Marrow stromal cells migrate throughout forebrain and cerebellum, and they differentiate into astrocytes after injection into neonatal mouse brains. Proc Natl Acad Sci USA *96:* 10711–10716.

Krause, D.S., N.D. Theise, M.I. Collector, O. Henegariu, S. Hwang, R. Gardner, S. Neutzel, S.J. Sharkis (2001) Multi-organ, multi-lineage engraftment by a single bone marrow-derived stem cell. Cell *105(3):* 369–77.

Kuhn H.G., C.N. Svendsen (1999) Origin, functions, and potential of adult neural stem cells. BioEssays *21:* 625–630.

Kuhn, H.G., H. Dickinson-Anson, F.H. Gage (1996) Neurogenesis in the dentate gyrus of the adult rat: age-related decrease of neuronal progenitor proliferation. J Neurosci *16:* 2027–2033.

Kuhn, H.G., J. Winkler, G. Kempermann, L.J. Thal, F.H. Gage (1997) Epidermal growth factor and fibroblast growth factor-2 have different effects of neural progenitors in the adult rat brain. J Neurosci *17:* 5820–5829.

Laywell, E.D., P. Rakic, V.G. Kukekov, E.C. Holland, D.A. Steindler (2000) Identification of a multipotent astrocytic stem cell in the immature and adult brain. Proc Natl Acad Sci USA *97:* 13883–13888.

Lim, D.A, A. Alvarez-Buylla (1999) Interaction between astrocytes and adult subventricular zone precursors stimulates neurogenesis. Proc Natl Acad Sci USA *96:* 7526–7531.

Lim, D.A., A.D. Tramontin, J.M. Trevejo, D.G. Herrera, J.M. Garcia-Verdugo, A. Alvarez-Buylla (2000) Noggin antagonizes BMP signalling to create a niche for adult neurogenesis. Neuron *26:* 713–726.

Loeffler, M., C.S. Potten (1997) Stem cells and cellular pedigrees – a conceptual introduction, in: Potten C.S. (ed.), Stem cells, Academic Press, London, pp. 1–27.

Lois, C., A. Alvarez-Buylla (1994) Long-distance neuronal migration in the adult mammalian brain. Science *264:* 1145–1148.

Lois, C., J.M. Garcia-Verdugo, A. Alvarez-Buylla (1996) Chain migration of neuronal precursors. Science *271(5251):* 978–981.

Luskin, M.B. (1993) Restricted proliferation and migration of postnatally generated neurons derived from the forebrain subventricular zone. Neuron *11:* 173–189.

Mabie, P.C., M.F. Mehler, R. Marmur, A. Papavasiliou, Q. Song, J.A. Kessler (1997) Bone morphogenetic proteins induce astroglial differentiation of oligodendroglial-astroglial progenitor cells. J Neurosci *17:* 4112–4120.

McKinney-Freeman, S.L., K.A. Jackson, F.D. Camargo, G. Ferrari, F. Mavilio, M.A. Goodell (2002) Muscle-derived hematopoietic stem cells are hematopoietic in origin. Proc Natl Acad Sci USA *99(3):* 1341–1346.

Mezey, E., K.J. Chandross, G. Harta, R.A. Maki, S.R. McKercher (2000) Turning blood into brain: cells bearing neuronal antigens generated in vivo from bone marrow. Science *290:* 1779–1782.

Mi, H., B.A. Barres (1999) Purification and characterization of astrocyte precursor cells in the developing rat optic nerve. J Neurosci *19:* 1049–1061.

Morrison, S.J., N.M. Shah, D.J. Anderson (1997) Regulatory mechanisms in stem cell biology. Cell *88:* 287–298.

Morshead C.M., B.A. Reynolds, C.G. Craig, M.W. McBurney, W.A. Staines, D. Morassutti, S. Weiss, D. van der Koy (1994) Neural stem cells in the adult mammalian forebrain: A relatively quiescent subpopulation of subependymal cells. Neuron *13:* 1071–1082.

Noctor, S.C., A.C. Flint, T.A. Weissman, R.S. Dammerman, A.R. Kriegstein (2001) Neurons derived from radial glial cells establish radial units in neocortex. Nature *409:* 714–720.

Palmer, T.D., E.A. Markakis, A.R.Willhoite, F. Safar, F.H. Gage (1999) Fibroblast growth factor-2 activates a latent neurogenic program in neural stem cells from diverse regions of the adult CNS. J Neurosci *198:* 487–8497.

Palmer, T.D., A.R. Willhoite, F.H. Gage (2000) Vascular niche for adult hippocampal neurogenesis. J Comp Neurol *425(4):* 479–494.

Pencea, V., K.D. Bingaman, L.J. Freedman, M.B. Luskin (2001a) Neurogenesis in the subventricular zone and rostral migratory stream of the neonatal and adult primate forebrain. Exp Neurol *172(1):* 1–16.

Pencea, V., K.D. Bingaman, S.J. Wiegand, M.B. Luskin (2001b) Infusion of brain-derived neurotrophic factor into the lateral ventricle of the adult rat leads to new neurons in the parenchyma of the striatum, septum, thalamus, and hypothalamus. J Neurosci *21(17):* 6706–6717.

Peretto, P., A. Merighi, A. Fasolo, L. Bonfanti (1999) The subependymal layer in rodents: a site of structural plasticity and cell migration in the adult mammalian brain. Brain Res Bull *49:* 221–243.

Petersen, B.E., W.C. Bowen, K.D. Patrene, W.M. Mars, A.K. Sullivan, N. Murase, S.S. Boggs, J.S. Greenberger, J.P. Goff (1999) Bone marrow as a potential source of hepatic oval cells. Science *284:* 1168–1171,

Pipia, G.G., H.P. Low, H.P.Turner, C. McAuliffe, R. Salmonsen, P.J.Quesenberry, W.J. Schwartz, N.S. Litofsky, A. Ross, L. Recht, S. Jones (2000) Transdifferentiation of neural precursor cells after blastocyst implantation. Society for Neuroscience's, 30th Annual Meeting, New Orleans, LA, November 4–9, 2000. Abstract n. 415.13, Vol. 1, p 1101.

Potten, C.S., M. Loeffler (1990) Stem cells: Attributes, cycles, spirals, pitfalls and uncertainties. Lesson for and from the crypt. Development *110:* 1001–1020.

Raff, M.C., L.E. Lillien, W.D. Richardson, J.F. Burne, M.D. Noble (1988) Platelet-derived growth factor from astrocytes drives the clock that times oligodendrocyte development in culture. Nature *333:* 562–565.

Reynolds, B.A., S. Weiss (1992) Generation of neurons and astrocytes from isolated cells of the adult mammalian central nervous system. Science *255:* 1707–1710.

Richards K.J., T.J. Kilpatrick, P.F. Bartlett (1992) De novo generation of neuronal cells from the adult mouse brain. Proc Natl Acad Sci USA *9:* 8591–8595.

Rooman, I., Y.Heremans, H. Heimberg, L. Bouwens (2000) Modulation of rat pancreatic acinoductal transdifferentiation and expression of PDX-1 in vitro. Diabetologia *43:* 907–914.

Sanchez-Ramos, J., S. Song, F. Cardozo-Pelaez, C. Hazzi, T. Stedeford, A Willing, T.B. Freema, S. Saporta, W. Janssen, N. Patel, D.R. Cooper, P.R. Sanberg (2000) Adult bone marrow stromal cells differentiate into neural cells in vitro. Exp Neurol *164:* 247–256.

Satho, J., Y. Kuroda (2000) Beta-catenin expression in human neural cell lines following exposure to cytokines and growth factors. Neuropathology *20:* 113–123.

Seale, P., L.A. Sabourin, A. Girgis-Gabardo, A. Mansouri, P. Gruss, M.A. Rudnicki (2000) Pax7 is required for the specification of myogenic satellite cells. Cell *102:* 777–786.

Seki, T., Y. Arai (1993) Highly polysialylated neural cell adhesion molecule (NCAM-H) is expressed by newly-generated granule cells in the dentate gyrus of the adult rat. J Neurosci *13:* 2351–2358.
Seri, B., J.M. Garcia-Verdugo, B.S. McEwen, A. Alvarez-Buylla (2001) Astrocytes give rise to new neurons in the adult mammalian hippocampus. J Neurosci *21:* 7153–7160.
Shihabuddin, L.S., P.J. Horner, J. Ray, F.H. Gage (2000) Adult spinal cord stem cells generate neurons after transplantation in the adult dentate gyrus. J Neurosci *20:* 8727–8735.
Shihabuddin, L.S., J. Ray, F.H. Gage (1997) FGF-2 is sufficient to isolate progenitors found in the adult mammalian spinal cord. Exp Neurol *148:* 577–586.
Slack, J.M.W. (1986) Epithelial metaplasia and the second anatomy. Lancet, August *2:* 268–270.
Suhonen, J.O., D.A. Peterson, J. Ray, F.H. Gage (1996) Differentiation of adult hippocampus-derived progenitors into olfactory neurons in vivo. Nature *383:* 624–627.
Temple, S. (1999) The obscure origins of adult stem cells. Curr Biol *9:* 397–399.
Temple, S., A. Alvarez-Buylla (1999) Stem cells in the adult mammalian central nervous system. Curr Op Neurobiol *9:* 135–141.
Theise, N.D., S. Badve, R. Saxena, O. Henegariu, S. Sell, J.M. Crawford, D.S. Krause (2000) Derivation of hepatocytes from bone marrow cells in mice after radiation-induced myeloablation. Hepatology *31:* 235–240.
Tropepe, V., C.G. Craig, C.M. Morshead, D. van der Kooy (1997) Transforming growth factor-alpha null and senescent mice show decreased neural progenitor cell proliferation in the forebrain subependyma. J Neurosci *17:* 7850–7859.
Uchida, N., D.W. Buck, D. He, M.J. Reitsma, M. Masek, T.V. Phan, A.S. Tsukamoto, F.H. Gage, I.L. Weissman (2000) Direct isolation of human central nervous system stem cells. Proc Natl Acad Sci USA *97:* 14720–14725.
Vescovi, A.L., B.A. Reynolds, D.D. Fraser, S. Weiss (1993) bFGF regulates the proliferative fate of unipotent (neuronal) and bipotent (neuronal/astroglial) EGF-generated CNS progenitor cells. Neuron *11:* 951–966.
Vicario-Abejon, C., K.K. Johe, T.G. Hazel, D. Collazo, R.D. McKay (1995) Functions of basic fibroblas growth factor and neurotrophins in the differentiation of hippocampal neurons. Neuron *15:* 105–114.
Wagner, J., P. Akerud, D.S. Castro, P.C. Holm, J.M. Canals, E.Y. Snyder, T. Perlmann, E. Arenas (1999) Induction of a midbrain dopaminergic phenotype in Nurr1-overexpressing neural stem cells by type 1 astrocytes. Nat Biotechnol *17:* 653–659.
Watt, F.M., B.L.M. Hogan (2000) Out of eden: stem cells and their niches. Science *287:* 1427–1430.
Weissmann, I.L. (2000) Stem cells: units of regenaration and units in evolution. Cell *100:* 157–168
Wilson, P.A., A. Hemmati-Brivanlou (1997) Vertebrate neural induction: inducers, inhibitors, and a new synthesis. Neuron *18:* 699–710.
Woodbury, D., E.J. Schwarz, D.J. Prockop, I.B. Black (2000) Adult rat and human bone marrow stromal cells differentiate into neurons. J Neurosci Res *61:* 364–370.
Yan, J., L. Studer, R.D. McKay (2001) Ascorbic acid increases the yield of dopaminergic neurons derived from basic fibroblast growth factor expanded mesencephalic precursors J Neurochem *76:* 307–311.
Zigova, T., V. Pencea, S.J. Wiegand, M.B. Luskin (1998) Intraventricular administration of BDNF increases the number of newly generated neurons in the adult olfactory bulb. Mol Cell Neurosci *11:* 234–245.

Cells
Tissues
Organs

Cells Tissues Organs 2002;171:77–89

Developmental Potential of Hematopoietic and Neural Stem Cells: Unique or All the Same?

Nicole Kirchhof[1] Friedrich Harder[1] Suzana Petrovic Simon Kreutzfeldt
Carolin Schmittwolf Michael Dürr Jan Kirsten Bettina Mühl
Angela Merkel Albrecht M. Müller

Institut für Medizinische Strahlenkunde und Zellforschung, Universität Würzburg, Germany

Key Words
Embryonic stem cells · Hematopoietic stem cells · Neural stem cells · Somatic stem cells · Plasticity

Abstract
Like many other animals, mammals develop from fertilized oocytes – the ultimate stem cells. As embryogenesis proceeds, most cells lose developmental potential and eventually become restricted to a specific cell lineage. The result is the formation of a complete and structured mature organism with complex organs composed of a great variety of mature, mostly mitotically quiescent effector cells. However, along the way, some exceptional cells, known as somatic stem cells (SSCs) are set aside and maintain a high proliferation and tissue-specific differentiation potential. SSCs, in contrast to embryonic stem (ES) cells, which are able to give rise to all cell types of the body, have been regarded as being more limited in their differentiation potential in the sense that they were thought to be committed exclusively to their tissue of origin. However, recent studies have demonstrated that somatic stem cells from a given tissue can also contribute to heterologous tissues and thus show a broad non-tissue restricted differentiation potential. The question arises: how plastic are somatic stem cells? To provide a tentative answer, we describe and review here recent investigations into the developmental potentials of two somatic stem cell types, namely hematopoietic and neural stem cells.

Introduction

The capacity of early embryonic cells becomes more and more restricted in the course of embryonic development. The apparently equivalent cells of the morula lose their totipotency during generation of the blastocyst stage

Abbreviations used in this paper

ES cells	embryonic stem cells
HSCs	hematopoietic stem cells
NSCs	neural stem cells
SSCs	somatic stem cells

[1] These authors contributed equally to this work.

KARGER
Fax + 41 61 306 12 34
E-Mail karger@karger.ch
www.karger.com

1422–6405/02/1711–0077$18.50/0

Accessible online at:
www.karger.com/journals/cto

Albrecht M. Müller
Institut für Medizinische Strahlenkunde und Zellforschung (MSZ), Universität Würzburg
Versbacherstrasse 5, D–97078 Würzburg (Germany)
Tel. +49 931 201 45848, Fax +49 931 201 45835
E-Mail albrecht.mueller@mail.uni-wuerzburg.de

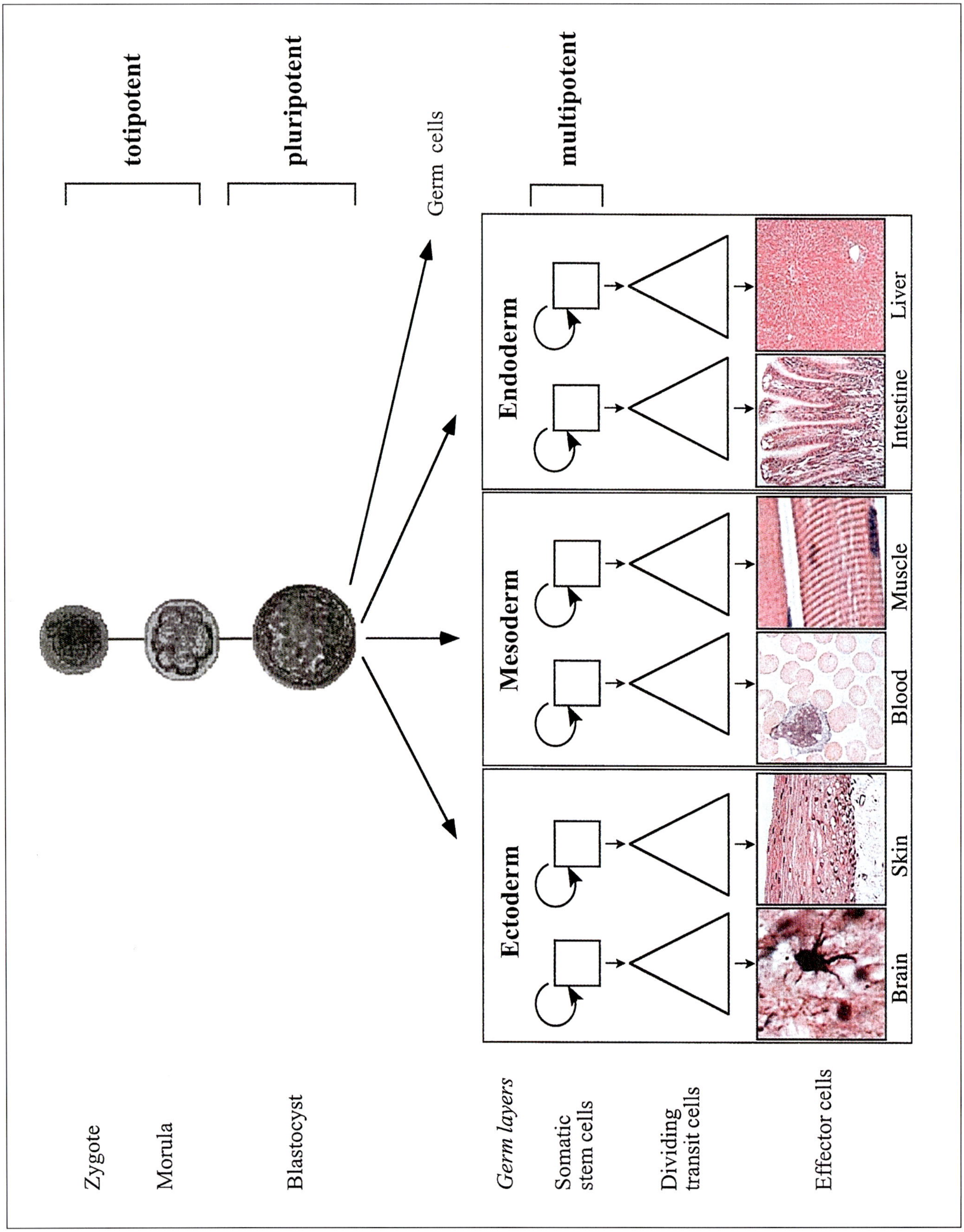
Zygote
Morula
Blastocyst
totipotent
pluripotent
Germ cells
multipotent
Germ layers
Somatic stem cells
Dividing transit cells
Effector cells
Ectoderm
Mesoderm
Endoderm
Brain
Skin
Blood
Muscle
Intestine
Liver

embryo giving rise to the first differentiated cells of the blastocyst. The blastocyst is composed of the trophectoderm and the inner cell mass, from which pluripotent embryonic stem cells can be derived. As gastrulation and cell division proceed, the cells of the inner cell mass differentiate further along various pathways, losing their pluripotency while generating all cell lineages and tissues of the developing embryo. Although most cells cease mitosis and proceed to terminal differentiation, some rare cells known as multipotent SSCs are set aside and maintain a tissue-specific differentiation potential. SSCs have been identified in a number of mammalian tissues (e.g. bone marrow, muscle, gut, skin, hair follicle, retina, peripheral and central nervous system [Fuchs and Segre, 2000]). They are the source of a variety of hierarchically structured cell systems with each type of stem cell at the top of a specific cellular differentiation cascade (fig. 1). SSCs represent a potent reservoir for cell replacement in tissues with high cell turnover due to natural cell death, injury or disease.

Stem cells are characterized by their potential to self-renew and to differentiate into several different lineages for the lifetime of the animal. In principle, stem cells can undergo two potential mitotic pathways: symmetric (both cells adopt the same fate) and asymmetric (daughter cell takes on different fate) cell division cycles. Observations of hematopoietic stem cell (HSC) cycling suggest that HSCs can go through asymmetric cell division cycles [Huang et al., 1999; Ema et al., 2000]. Both types of cell divisions have been observed in time-lapse video-microscopy in murine central nervous system (CNS) stem cells [Qian et al., 1998]. Asymmetric cell division of a stem cell generates cells with different fates and results in the generation of a new stem cell and a less primitive progenitor cell. Progenitor cells, in contrast to stem cells, have only limited self-renewal and limited multi-lineage differentiation potential. Finally, progenitor cells proliferate and mature further to produce many differentiated effector cells that have no self-renewal and very restricted differentiation potential [Loeffler and Potten, 1997]. It should be emphasized that this model was established analyzing stem and progenitor cell relationships in the intestinal epithelium and in the hematopoietic system and it is therefore at present unclear, whether it applies to stem cells at all developmental stages and to all SSC systems in the adult organism. Stem cell systems have been observed in many metazoan animals including C. elegans, Drosophila and newt, and it is conceivable that comparative studies may provide further insights for a better understanding of mammalian stem cells.

Fig. 1. Hierarchy and potential of stem cells during ontogeny. During the course of mammalian embryonic development, the zygote gives rise to the morula- and blastocyst-stage embryo, which in a process called gastrulation generates the three germ layers (ecto-, meso-, and endoderm). The embryo matures further through organogenesis and growth. But not all cell divisions generate terminally differentiated cells, some special cells are retained as SSCs maintaining a high proliferation and differentiation potential. Throughout ontogeny, stem cells with characteristic developmental potentials are generated (totipotent, pluripotent and multipotent stem cells). All somatic cells are ultimately lost, with the death of the individual, whereas germ cells are set aside and give rise to the next generation [Pesce et al., 1999].

In mammals, isolated SSCs from adult tissues and organs were regarded as restricted to their tissue of origin, but more recent and surprising findings demonstrate that the developmental potential of SSCs is not restricted to only one tissue. An example of the plasticity within one stem cell system is the activation of an embryonic-type gene transcription program in progeny of adult-type HSCs by embryonic environments [Geiger et al., 1998]. Examples of even broader SSC plasticity are given by recent reports that cultured NSCs and stem cells in the muscle (also called side-population (SP) cells as determined by Hoechst 33342 dye exclusion) have hematopoietic activity and that isolated HSCs undergo fate changes and develop into muscle, brain or liver [Bjornson et al., 1999; Gussoni et al., 1999; Lagasse et al., 2000; Mezey et al., 2000]. Even multi-organ and multi-lineage engraftment has been observed for a single bone marrow-derived HSC [Krause et al., 2001]. These observations indicate that lineage conversion may occur between stem cells originally derived from different embryonic germ layers such as neural and hematopoietic cells. Furthermore, it has been demonstrated that even unipotent oligodendrocyte precursors can revert to multipotent central nervous system stem cells and that differentiated hematopoietic cells such as B lymphoid cells are not irreversibly committed and can switch to the myeloid cell lineage [Klinken et al., 1988; Principato et al., 1990; Kondo and Raff, 2000]. This apparent plasticity in somatic stem and committed cells opens the question of how primitive these cell populations are and whether SSCs carry tissue-specific properties at all.

To investigate and compare the developmental potential of two different SSC types we exposed HSCs and NSCs to an environment that has inductive character and is permissive for the development of all cell types of the embryo, the preimplantation blastocyst. This experimental system represents a powerful tool to further investigate the differentiation potential of different types of SSCs. Our studies have revealed both, a degree of tissue-specific identity and the presence of some plasticity within neural and hematopoietic stem cells, suggesting that these two SSC types have distinct capabilities but that they can also contribute to heterologous cell systems. Knowledge about the regulation of tissue-specific identities and plasticity will provide tools to utilize SSCs in autologous tissue replacement.

Isolation and Stem Cell Activity of Hematopoietic and Neural Stem Cells

As plasticity is attributed only to cells at the progenitor or multipotent stem cell level, the isolation of progenitor and stem cells is an important prerequisite for the characterization and further understanding of the phenomenon of stem cell plasticity. HSCs like other SSCs are a rare cell type. Estimates of HSC frequencies in fetal liver and bone marrow analyzed in competitive repopulation assays revealed that about 1 to 10 per 10^5 cells in the fetal liver and in the adult bone marrow are functional HSCs [Harrison et al., 1989; Szilvassy et al., 1990; Rebel et al., 1996; Zhong et al., 1996].

Enrichment and isolation of HSCs can principally be done by various means such as labelling with monoclonal antibodies against phenotypic surface markers and sorting using flow cytometry [Spangrude et al., 1988; Okada et al., 1992], by taking advantage of distinctive physical properties such as buoyant density [Turner et al., 1966; Haskill and Moore, 1970; Trevisan and Iscove, 1995], by using stem cell-specific dye efflux properties [Goodell et al., 1997], or by using a combination of several purification strategies. Murine HSCs have been purified from the fetal liver [Jordan et al., 1990] and the adult bone marrow [Visser et al., 1984; Lord and Spooncer, 1986; Spangrude et al., 1988]. They are characterized by the absence or low level expression of cell surface markers present on mature and committed hematopoietic cells and by high expression of both, the receptor tyrosine kinase c-kit and the stem cell antigen-1 (Sca-1), a phosphatidylinositol-anchored cell surface protein [Okada et al., 1992]. This cell population contains all of the HSC activity found in the bone marrow of mice [Uchida and Weissman, 1992]. Corresponding antibodies are used for the purification of human HSCs [Bhatia et al., 1997; Conneally et al., 1997]. One robust method of HSC purification from murine bone marrow therefore uses a combined negative and positive selection strategy: first all bone marrow cells that express the lineage marker epitopes TER119 (erythroid), Mac-1 (macrophage), Gr-1 (granulocytic), CD4 and CD8 (T lymphoid) and B220 (B lymphoid) are depleted, before c-kit and Sca-1 expressing cells are sorted using a flow cytometer. This procedure purifies Lin^-, $Sca\text{-}1^+$, $c\text{-}kit^+$ cells that represent a highly enriched HSC population [Okada et al., 1992]. A 6- to 8-week-old C57Bl/6 mouse has about 3×10^8 cells in the bone marrow; of these, 1–2% are Lin^- cells and approximately 0.01–0.02% of total bone marrow cells are of HSC phenotype [Ikuta et al., 1990; Morrison and Weissman, 1994; Jordan et al., 1995]. Limiting dilution analysis of Lin^-, $Sca\text{-}1^+$, $c\text{-}kit^+$ cells shows that this cell population is capable of long-term and multi-lineage hematopoietic reconstitution of lethally irradiated mice [Okada et al., 1992]. The principle of stem cell self-renewal was established by demonstration of their ability to repopulate the entire hematopoietic system in serial transplantations [Wu et al., 1968; Keller and Snodgrass, 1990].

Despite long-term and intensive research, isolation and ex vivo culture of HSCs thus far has led to a rapid loss of stem cell activity due to differentiation and apoptosis [Traycoff et al., 1998]. Experiments with transgenic mice carrying the human bcl-2 transgene under the combined transcriptional control of the mouse $H2K^b$ promoter and the Moloney murine leukemia virus LTR showed an unexpected modest role of bcl-2 in affecting the differentiation and proliferation characteristics of stem cells [Akashi et al., 1997; Kondo et al., 1997], and further analysis of H2K-BCL-2 transgenic mice revealed a limited increase in phenotypic HSC numbers with a slight advantage in competitive repopulation compared to wild-type HSCs [Domen et al., 2000]. The intravenous injection of only 20 Lin^-, $Sca\text{-}1^+$, $c\text{-}kit^+$ cells of non-transgenic or H2K-BCL-2 transgenic donor animals into irradiated recipients is sufficient to long-term repopulate the hematopoietic system (fig. 2). HSCs from H2K-BCL-2 transgenic mice are protected against apoptosis induced by growth factor deprivation and are more radio-resistant than HSCs of non-transgenic mice [Domen et al., 1998]. Therefore, these mice provide more robust HSCs for heterochronic transplantation settings such as injections into early embryos than HSCs from wild-type mice. Accordingly, we observed that following blastocyst injection of H2K-BCL-2 transgenic HSCs an increase in the level of male donor cells in the chimeric animals as compared to experiments in which HSCs of non-transgenic donors were used. Importantly, no changes in the tissue-specific repopulation

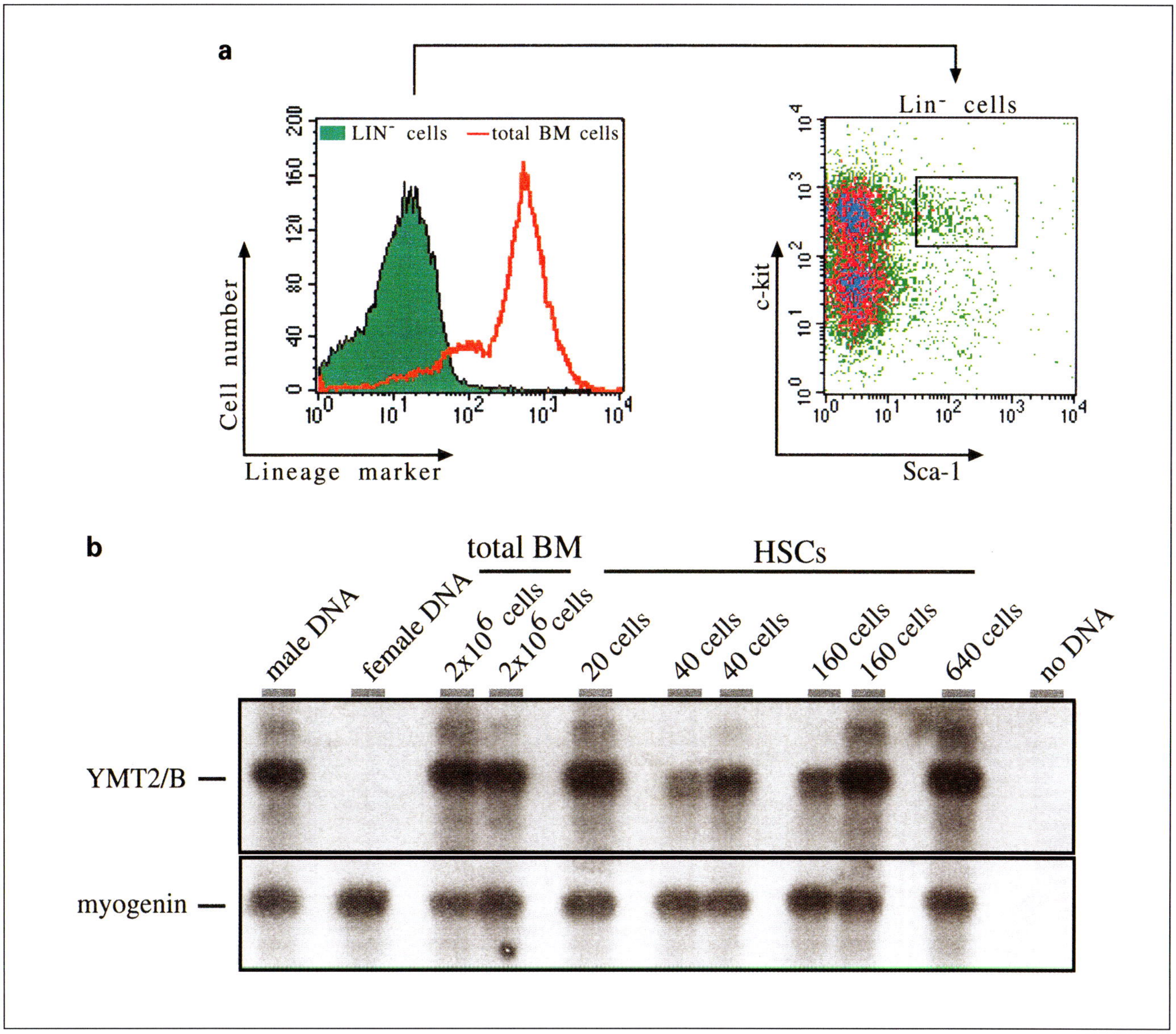

Fig. 2. Isolation of bone marrow-derived hematopoietic stem cells. **a** Whole bone marrow cells of 6–10 weeks old human BCL-2 transgenic mice [Domen et al., 1998] were isolated and fractionated by flow cytometry. Shown are the result of FACS analysis of whole bone marrow cells (red) and isolated Lin⁻ cells (green). Lin⁻ cells are further fractionated with the monoclonal antibodies anti-c-kit and anti-Sca-1. HSCs are highly enriched in the Lin⁻, Sca-1, c-kit⁺ cell fraction (gated region) [Okada et al., 1992]. **b** Limiting numbers of Lin⁻, Sca-1⁺, c-kit⁺ HSCs from male donor animals were injected into lethally irradiated female recipients and donor contribution was determined in the peripheral blood 4 months post transplantation by male-(YMT2/B) and myogenin-specific PCR [Geiger et al., 1998; Müller and Dzierzak, 1993]. Shown is an autoradiograph of a Southern blot. Abbreviations: BM = bone marrow; HSCs = hematopoietic stem cells; Lin = lineage cocktail.

patterns of transgenic and non-transgenic HSCs were observed (fig. 3; table 1).

The notion that multipotent NSCs exist emerged only recently. Although retroviral marking experiments suggested the presence of multipotent cells that can give rise to both neurons and glia in the brain [Price et al., 1987; Reid et al., 1995; Williams and Price, 1995], the identification of EGF and bFGF as growth factors for neural precursors established the presence of stem cells in the fetal and adult brain. NSCs, like other stem cells, are charac-

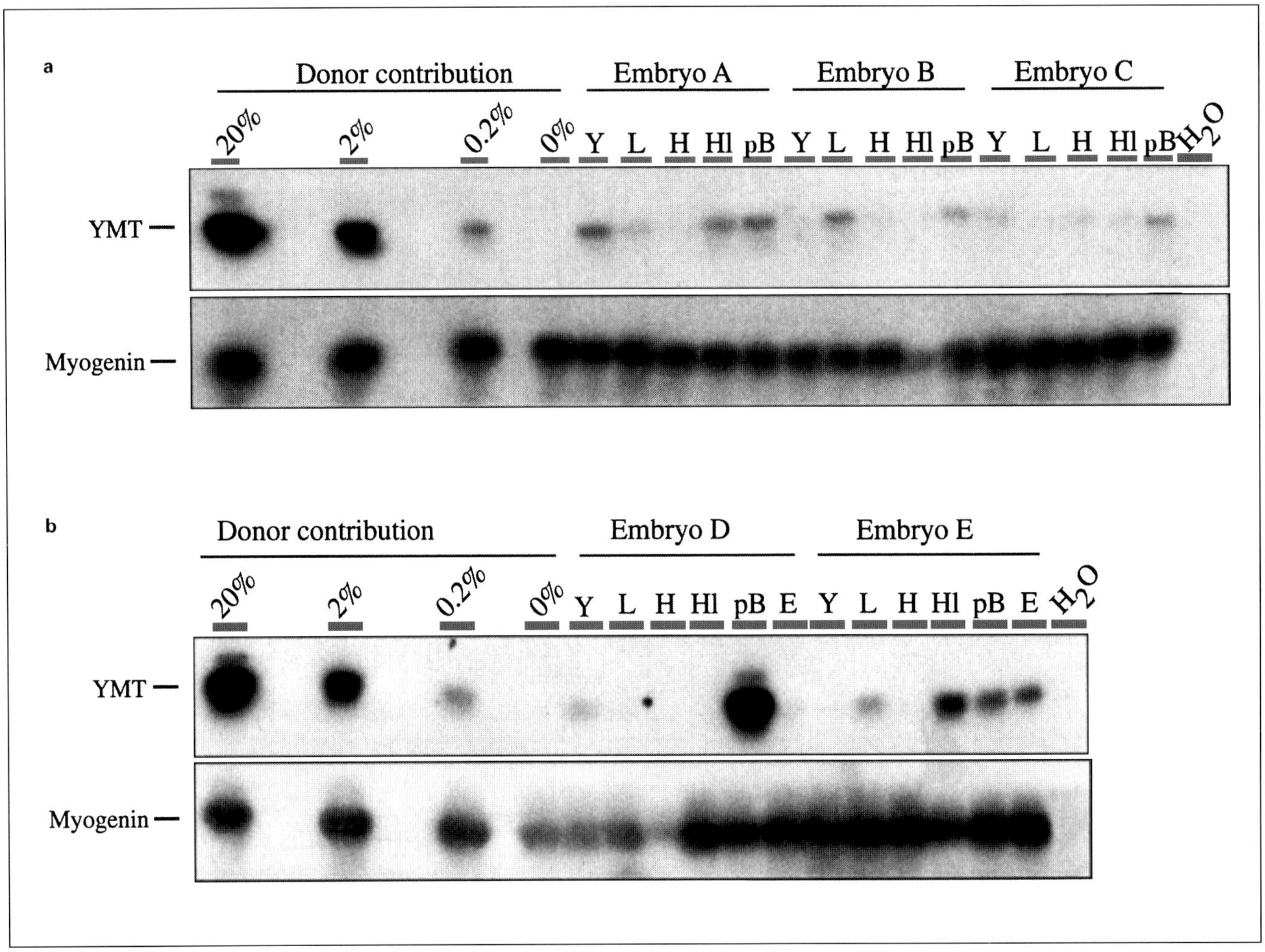

Fig. 3. Repopulation pattern of wt and hu-bcl2 transgenic HSCs following blastocyst injection. Murine HSCs derived from (**a**) wildtype and (**b**) BCL-2 transgenic male animals were injected and donor contribution was determined using YMT2/B-specific primers for PCR on total genomic DNA prepared from female E11.5 embryos. Dilutions of male into female genomic DNA were used for quantification of PCR products. Abbreviations: Y = yolk sac; L = fetal liver; H = head; Hl = hind limb; pB = peripheral blood. Autoradiographs of Southern blots are shown.

Table 1. Frequency of donor contribution following injection of wt and H2K-BCL-2 transgenic HSCs

	Tissues analyzed (positive/total)			
	Y	L	H	pB
wt HSC				
E11.5	7/8	8/8	4/17	5/10
E12.5	3/9	2/9	2/9	3/5
E13.5	6/7	2/7	0/7	2/3
bcl2-tg HSC				
E12.5	10/14	9/14	0/12	11/12

terized by the presence of two activities: self-renewal and multi-lineage differentiation. NSCs which can be expanded in vitro have been identified in embryonic day 14 (E14) mouse primordial striatum. The EGF-responsive embryonic progenitors express nestin, an intermediate filament found in neuroepithelial stem cells and in vitro produce neurons and glial cells [Reynolds and Weiss, 1992]. NSCs, which remain multipotent in culture, can also be isolated from the embryonic rat cerebral cortex [Davis and Temple, 1994]. Apart from multipotent progenitors there are also uni- and bipotent progenitors present in the murine E14 forebrain [Vescovi et al., 1993].

In the adult, NSCs are detected in a variety of brain structures such as the striatum, the subventricular zone, the dentate gyrus of the hippocampus, and in the spinal cord [Reynolds and Weiss, 1992; Morshead et al., 1994; Weiss et al., 1996; Gage, 2000]. Thus, NSCs are present at many different places in the developing and adult nervous system.

To demonstrate neural stem cell activity of E14 or E15 forebrain-derived neurosphere cultures, we tested whether our cultures exhibit properties that comply with the criteria representative of stem cells (fig. 4). Neurospheres were generated from wild-type and from H2K-BCL-2 transgenic embryos. The two populations showed no significant differences in plating efficiencies or differentiation abilities (data not shown). The number of spheres generated from E14 mouse striata was previously estimated to be 52 per 5,000 viable cells [Reynolds and Weiss, 1996]. In a first set of experiments we tested the in vitro differentiation capacities of expanded NSCs by differentiating single cell suspensions of neurospheres on poly-*L*-ornithine-coated surfaces under low serum conditions and in the absence of growth factors [Ahmed et al., 1995]. Under these conditions neurosphere-derived cells show little proliferation and differentiate rapidly within a few days. As shown in figure 4b, after four days of in vitro differentiation NSCs generate differentiated progeny which display distinctive neuronal and glial morphologies and are immuno-reactive with neuronal- and glial-specific monoclonal antibodies. The multi-lineage differentiation capacity is found in multi-clonal cultures and at the single NSC level (fig. 4b). This demonstrates that EGF- and bFGF-responsive NSCs are able to generate neurons, astrocytes and oligodendrocytes, the three main cell types of the CNS. Furthermore, transplantations of cultured NSCs into the brain of recipient animals showed engraftment, migration and differentiation to glial and neuronal progeny in a manner appropriate to the site of engraftment [Lois and Alvarez-Buylla, 1994; Suhonen et al., 1996].

The self-renewal potential of neurospheres was tested by serially diluting single cell suspensions generated by trypsin treatment of in vitro expanded neurospheres. After 2 weeks of culture in the presence of EGF and bFGF, the number of newly generated secondary spheres was plotted against the number of cells seeded per well. The cumulative data from six independent experiments is given in figure 4c. We find by linear regression analysis that single cell suspensions of neurospheres contain on average 3.8% cells with neurosphere-initiating activity. Thus, one neurosphere-initiating cell is present per 26 cells in our cultures. This frequency of neurosphere-forming cells is in agreement with previously published data [Reynolds and Weiss, 1996] and taken together with their multi-lineage differentiation activities shows that single cells derived from neurospheres are bona fide stem cells.

Developmental Potential of Hematopoietic and Neural Stem Cells

SSCs isolated from differentiated tissues were regarded in the past as being limited as to the cell types they can produce. Such as, that HSCs generate only blood cells. The tissue-specific identity of HSCs was first recognized when it was found that the effects of fatal radiation syndrome could only be prevented either by shielding the spleen or long bones, or by injecting bone marrow suspensions into irradiated hosts [Jacobson et al., 1951; Lorenz et al., 1951]. This concept that SSCs were limited to specific tissues was supported by analysis of the origin of the location of HSCs during fetal development which revealed hematopoietic activity first in the AGM (aorta gonad mesonephros) region, and later during development in the fetal liver and bone marrow, but not in somites, lung, heart or head [Medvinsky et al., 1993; Müller et al., 1994; Medvinsky et al., 1996]. But more recently, this concept has been challenged by a series of observations in which SSCs actually exhibited further differentiation activities. This raises the question of how broad is the differentiation potential of SSCs? Are different SSCs types capable of generating all cell types? Such a possibility was suggested in a report of J. Frisen and colleagues, who provided evidence that neurospheres can develop into many non-neural cell types [Clarke et al., 2000] implying that some SSCs such as NSCs have differentiation capabilities similar to those of pluripotent ES cells. To analyze the differentiation capacities of different SSC types and to investigate whether they carry tissue-specific identities, we transplanted about 40 NSCs, or HSCs, isolated from male donor animals into murine blastocysts. NSCs and HSCs were from H2K-BCL-2 transgenic mice. Previous experiments showed that HSCs isolated from H2K-BCL-2 transgenic mice are protected against apoptotic stimuli. Analysis of the tissue distribution of male donor cells in developing female animals was done by Y chromosome-specific PCR (fig. 5). Following HSC injection using H2K-BCL-2 transgenic donor animals into blastocysts, we detected donor chimerism in 8 of 23 (35%) adult animals. Whereas following NSCs

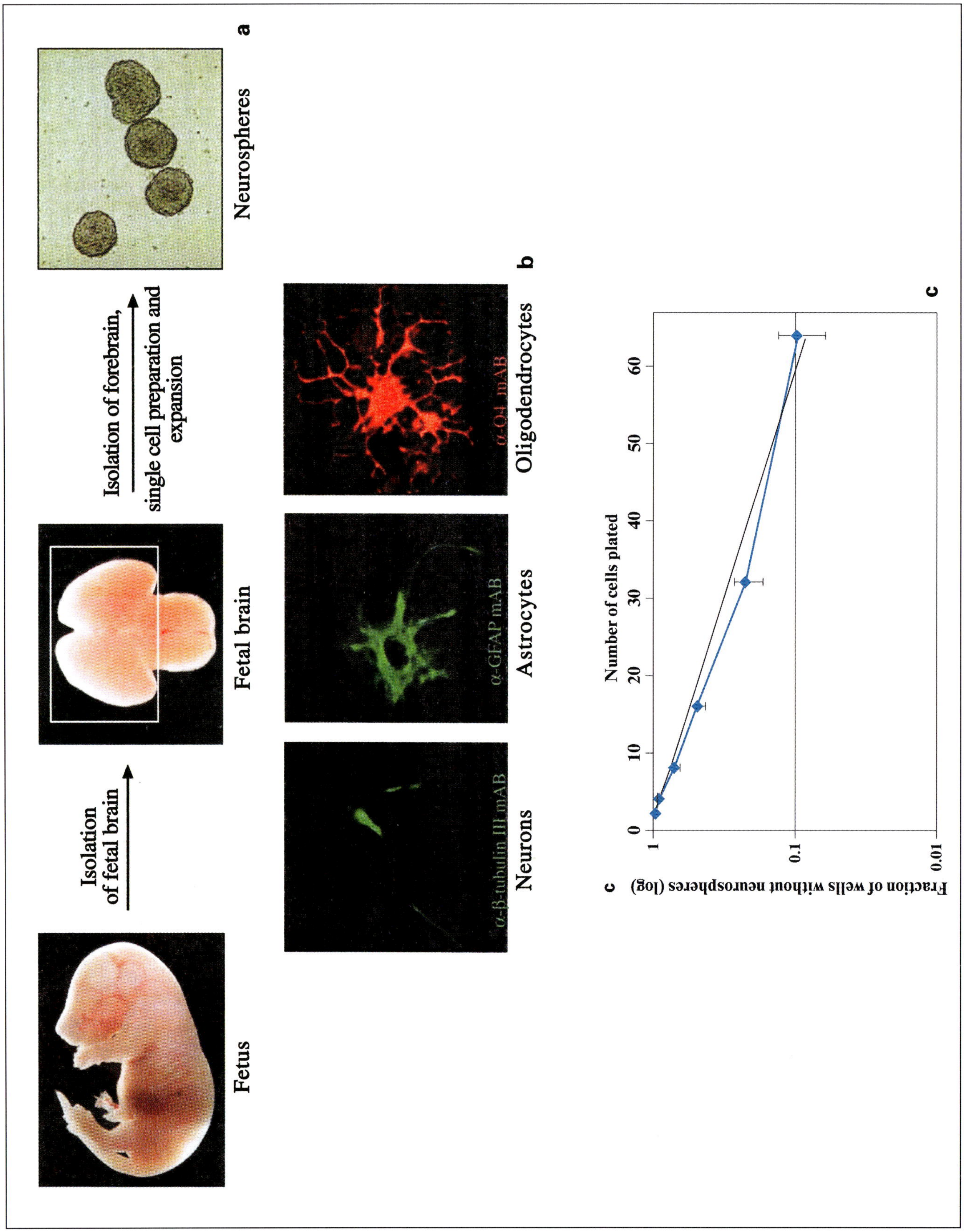
Fetus
Isolation
of fetal brain
Fetal brain
Isolation of forebrain,
single cell preparation and
expansion
Neurospheres
a
α-β-tubulin III mAB
Neurons
α-GFAP mAB
Astrocytes
α-O4 mAB
Oligodendrocytes
b
Number of cells plated
Fraction of wells without neurospheres (log)
0
10
20
30
40
50
60
1
0.1
0.01
c

4

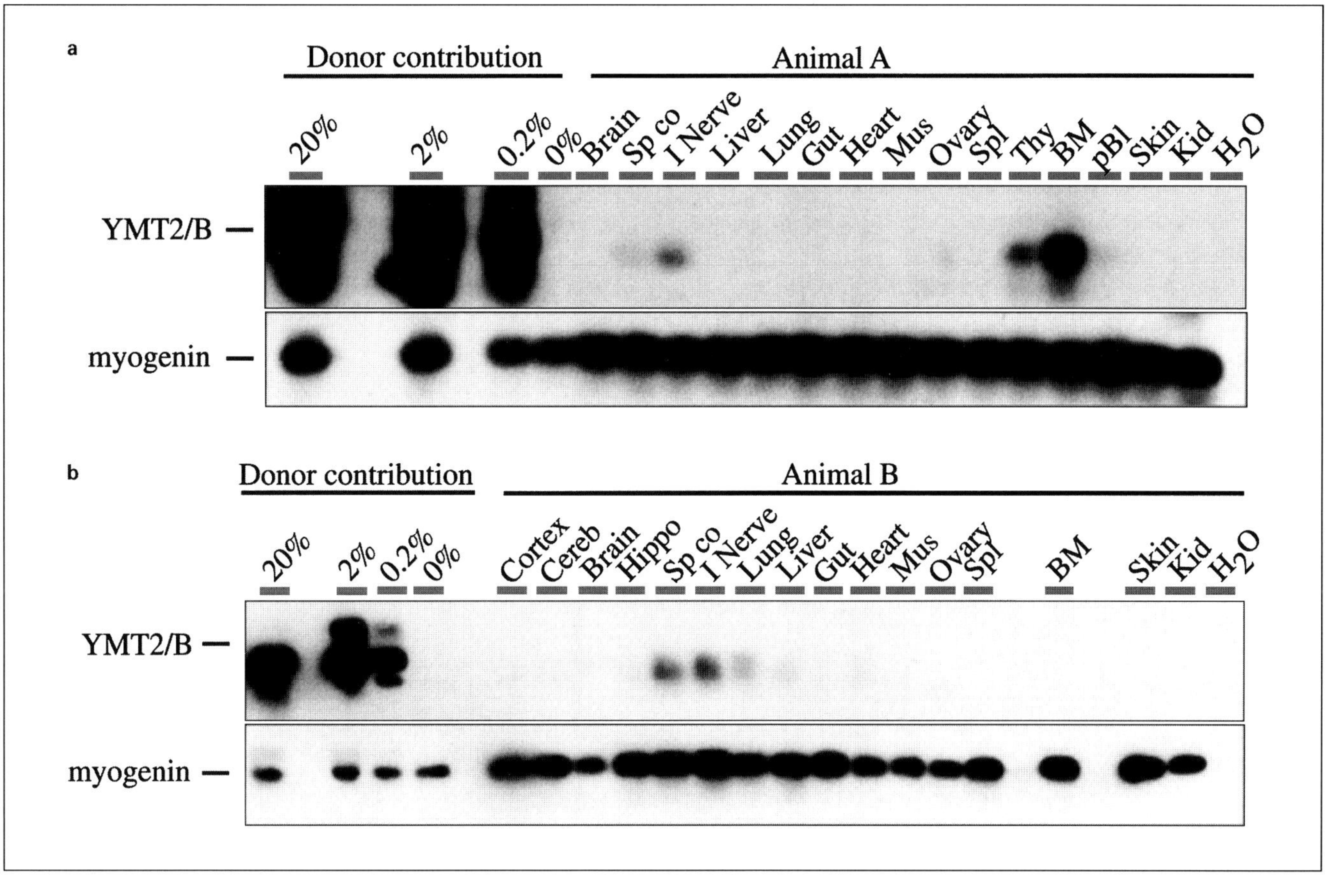

5

Fig. 4. Isolation and culture of fetal brain-derived NSCs. **a** For the isolation of NSCs, forebrains of individual BCL-2 transgenic embryos at E14 or E15 were dissected, single cell suspensions were prepared by trypsinization and cells were expanded in neural basal medium supplemented with B27 supplement (Gibco/BRL) in the presence of EGF (20–40 ng/ml) and bFGF (20–40 ng/ml) [Reynolds and Weiss, 1992; Vescovi and Snyder, 1999]. NSCs were propagated as floating aggregates called neurospheres in tissue culture flasks with no substrate pretreatment. **b** Single cell suspensions of neurospheres upon cultivating onto poly-*L*-ornithine-coated plates under differentiation conditions (1% serum and absence of EGF, bFGF), produced three neural cell types: neurons (beta-tubulin III), oligodendrocytes (O4) and astrocytes (GFAP: glial fibrillaric acidic protein). **c** To assess the ability of neurospheres to self-renew, single cell suspensions were prepared from neurospheres by trypsinization (passage number >6) and graded numbers of cells were seeded into 96-well plates. Wells that developed no neurospheres 2 weeks post seeding were scored as negative and fractions of negative wells were plotted against the numbers of cells seeded. Results from 6 different analyses were combined and error bars (SEM) are shown. The linear regression curve ($y = e^{-0.0387x}$) was calculated with the statistics function of the Microsoft Excel 2000 software and the function was used to calculate the frequency of NSCs.

Fig. 5. Donor contribution in chimeric recipients upon injection of HSCs and NSCs into blastocysts. **a** Bone marrow-derived HSCs from male animals were injected into blastocysts and developing females were sacrificed at 4 months of age. Genomic DNA was prepared from a female recipient and male donor contribution was assessed. **b** In vitro cultured and expanded NSCs of male donor fetuses were dissociated into a single cell suspension and following one day of growth were injected as small spheres into a blastocyst. Donor contribution in a female chimeric recipient at the age of 4 months was determined by male-specific PCR as described in figure 2. Abbreviations: Sp Co = spinal cord; I Nerve = ischiadic nerve; Mus = muscle; Spl = spleen; Thy = thymus; BM = bone marrow; pBI = peripheral blood; Kid = kidney; Cereb = cerebellum; Hippo = hippocampus.

injection of transgenic mice into blastocysts we detected donor chimerism in 6 of 13 (46%) adult animals. Analysis of the donor contribution in animal A, which was 4 months of age and developed following the injection of bone marrow-derived HSCs, revealed a significant number of male cells in the hematopoietic tissues thymus and bone marrow and somewhat lower numbers in spinal cord and ischiadic nerve. This pattern of contribution with higher signals in hematopoietic tissues and less signal in other tissues is typical for animals developing after HSC injection.

Functional donor-derived hematopoietic progenitor cells exist in chimeric hematopoietic tissues after injection of neomycin resistant transgenic HSCs into blastocysts, as shown by the outgrowth of hematopoietic colonies in the presence of appropriate growth factors and G418 [Geiger et al., 1998]. Currently, the cellular nature of donor cells in non-hematopoietic tissue is unclear, but since the circulation of animal A is devoid of donor cells it is unlikely that the signals originate from blood.

NSCs are another source of SSCs capable of generating adult chimeric recipients following injection into blastocysts. Analysis of donor contribution revealed NSC-derived chimeric tissues in a 4-month-old recipient. The male donor cells were mainly detected in the neural tissues spinal cord and ischiadic nerve and only trace amounts in liver (animal B, fig. 5b). Furthermore, after the injection of NSCs isolated from transgenic animals expressing the lacZ marker ubiquitously (HMG CoA promoter controlling the lacZ transgene) we could detect lacZ positive cells the Schwann cells surrounding the ischiadic nerve of an adult animal [Harder et al., in preparation]. Schwann cells are neural crest cell derived and contribution to cells of this lineage shows that CNS-derived NSCs can produce peripheral nervous system cells. Contribution to non-neural tissues was detected by injecting NSCs from human β-globin gene locus transgenic mice [Strouboulis et al., 1992]. The human β-globin transgene locus has been shown to be expressed in mice in a tissue-specific and developmental stage-specific manner in erythroid cells only, suggesting the generation of NSC-derived erythroid cells [Müller et al., 2000; Harder et al., in preparation]. As shown in table 2, HSCs and NSCs show distinct tissue-specific repopulation preferences. Analysis of the tissue distribution and the number of donor cells in chimeric tissues revealed that NSCs preferentially produce a neural-specific pattern of tissue engraftment, whereas HSCs are preferentially detected in hematopoietic and at a lower frequency and intensity in non-hematopoietic tissues. These results indicate, that even if SSCs from different tissue origins are injected into an environment which is permissive for the development of all cell lineages of the embryo, they predominantly retain their tissue-specific identity.

Table 2. Analysis of adult animals developing from blastocysts injected with HSCs or NSCs

Stem cell type	Tissue analyzed (positive/total)				
	Cortex	Sp.c.	Spleen	Thymus	BM
HSCs	1/23	5/23	2/23	7/23	3/23
NSCs	3/13	6/13	0/13	1/11	0/11

Sp.c. = spinal cord; BM = bone marrow.

Summary and Discussion

The data described in this report support the conservative hypothesis that neural and hematopoietic stem cells predominantly behave as tissue-specific stem cells. However, despite their tissue specificity, they show a certain degree of broader differentiation potential as seen in their ability to contribute to heterologous tissues. The tissue specificity of SSCs is no surprise, as each stem cell has its own unique developmental history and role. Furthermore, each SSC type resides in a complex and different environment, the stem cell niche. There is evidence that the stromal and stem cell components mutually regulate each other during development [Hollander et al., 1995]. In the stem cell niche, multiple tissue-specific stem cell stroma interactions occur via integral membrane proteins and the extracellular matrix and also through secreted factors that in concert regulate the fate of stem cells and their progeny [Whetton and Graham, 1999; Watt and Hogan, 2000]. The existence of tissue-specific stem cell niches also predicts that different stem cell types rely on distinct cell-cell contacts and growth factor signals for survival, self-renewal and differentiation. Indeed, ES cells and NSCs require different growth factors and cytokines for survival and self-renewal in vitro and despite years of intensive research, HSCs still cannot be readily maintained and expanded in vitro. Thus, ES cells and SSCs isolated from differentiated tissues are distinct cellular entities and the repopulation preferences of HSCs and NSCs in an environment which is open for the development of all cell lineages reflects a predominant commitment due to the origin of the somatic stem cells.

Nevertheless mammalian stem cells (or a subset of those cells) of one adult tissue have some potential to contribute to a variety of other cell systems. The cellular basis for such rare events is currently elusive. Several hypotheses are tenable: It is for example conceivable, that SSCs revert to an ES cell-like developmental state before generating heterologous cell systems. Alternatively, somatic stem cells might directly transform into somatic stem cells of other cell systems. A third hypothesis would predict that certain SSCs have bi- or multipotential activities such as neuro-hematopoietic and neuro-muscular activities but are not pluripotent. Distinguishing between these options requires clonal examinations whereby the cellular progeny of a single stem cell is analyzed.

Of interest to laboratories studying stem cell plasticity are the results of microarray-based gene expression analysis of NSCs and HSCs which showed that differences and overlaps exist. It is possible that unidentified gene products, expressed in several different stem cell types, are the cause for the apparent overlap of the developmental capacities of different SSCs [Geschwind et al., 2001].

While the lifelong activity of SSCs in mammals to maintain tissue homeostasis is impressive, the apparent developmental potentials of mammalian SSCs seem weak, compared to complete limb regeneration in newt, or transdetermination of imaginal discs in Drosophila. Nevertheless, the phenomena of SSCs contributing to cell types of different germ layer origin might be similar to processes described in Drosophila [Wei et al, 2000]. The Drosophila imaginal discs, the anlagen of adult cuticular structures such as antennae, legs and wings, are generated during embryogenesis and are committed to give rise to one structure. Transplantation experiments have shown that this disc-specific determination is faithfully maintained. In rare cases imaginal discs can be made to switch their determined state, for example leg disc cells can be made to develop into a wing. This process is known as transdetermination [Hadorn, 1968]. The results of transdeterminations resemble the effect of homeotic mutations in which one body or organ segment due to a mutation in a Hox gene is transformed into a different but perfectly normal segment [Schneuwly et al, 1987]. Interestingly, vertebrate Hox genes are not only involved in patterning of the embryo but also play a role in blood cell differentiation [Magli et al., 1997]. Parallels between transdetermination in Drosophila and mammalian stem cell plasticity suggest that both phenomena rely on primitive cells with a high proliferative capacity and might be based on similar molecular mechanisms conserved throughout evolution. Understanding the physiological processes underlying fate changes in non-mammalian organisms will provide clues for the manipulation of mammalian stem cells.

Acknowledgments

The authors are indebted to Randy Cassada, Hartmut Geiger, Gerald de Haan and Joe Carnwath for critically reading the manuscript; Michael Sendtner and Stefan Wiese for help with NSC cultures; Sonja Rotzoll for expert assistance with flow cytometry; and Jos Domen, Irving Weissman and Frank Grosveld for providing transgenic animals. This work was supported by the Deutsche Forschungsgemeinschaft (SFB 465 and SP 1109).

References

Ahmed, S., B.A. Reynolds, S. Weiss (1995) BDNF enhances the differentiation but not the survival of CNS stem cell-derived neuronal precursors. J Neurosci *15:* 5765–5778.

Akashi, K., M. Kondo, U. von Freeden-Jeffry, R. Murray, I.L. Weissman (1997) Bcl-2 rescues T lymphopoiesis in interleukin-7 receptor-deficient mice. Cell *89:* 1033–1041.

Bhatia, M., J.C.Y. Wang, U. Kapp, D. Bonnet, J.E. Dick (1997) Purification of primitive human hematopoietic cells capable of repopulating immune-deficient mice. Proc Natl Acad Sci USA *94:* 5320–5325.

Bjornson, C.R.R., R.L. Rietze, B.A. Reynolds, M.C. Magli, A.L. Vescovi (1999) Turning brain into blood: a hematopoietic fate adopted by adult neural stem cells in vivo. Science *283:* 534–537.

Clarke, D.L., C.B. Johansson, J. Wilbertz, B. Veress, E. Nilsson, H. Karlstrom, U. Lendahl, J. Frisen (2000) Generalized potential of adult neural stem cells. Science *288:* 1660–1663.

Conneally, E., J. Cashman, A. Petzer, C. Eaves (1997) Expansion in vitro of transplantable human cord blood stem cells demonstrated using a quantitative assay of their lympho-myeloid repopulating activity in nonobese diabetic-scid/scid mice. Proc Natl Acad Sci USA *94:* 9836–9841.

Davis, A.A., S. Temple (1994) A self-renewing multipotential stem cell in embryonic rat cerebral cortex. Nature *372:* 263–266.

Domen, J., S.H. Cheshier, I.L. Weissman (2000) The role of apoptosis in the regulation of hematopoietic stem cells: Overexpression of Bcl-2 increases both their number and repopulation potential. J Exp Med *191:* 253–263.

Domen, J., K.L. Gandy, I.L. Weissman (1998) Systemic overexpression of BCL-2 in the hematopoietic system protects transgenic mice from the consequences of lethal irradiation. Blood *91:* 2272–2282.

Ema, H., H. Takano, K. Sudo, H. Nakauchi (2000) In vitro self-renewal division of hematopoietic stem cells. J Exp Med *192:* 1281–1288.

Fuchs, E., J.A. Segre (2000) Stem cells: a new lease on life. Cell *100:* 143–155.

Gage, F.H. (2000) Mammalian neural stem cells. Science *287:* 1433–1438.

Geiger, H., S. Sick, C. Bonifer, A.M. Müller (1998) Globin gene expression is reprogrammed in chimeras generated by injecting adult hematopoietic stem cells into mouse blastocysts. Cell *93:* 1055–1065.
Geschwind, D.H., J. Ou, M.C. Easterday, J.D. Dougherty, R.L. Jackson, Z. Chen, H. Antoine, A. Terskikh, I.L. Weissman, S.F. Nelson, H.I. Kornblum (2001) A genetic analysis of neural progenitor differentiation. Neuron *29:* 325–339.
Goodell, M.A., M. Rosenzweig, H. Kim, D.F. Marks, M. DeMaria, G. Paradis, S.A. Grupp, C.A. Sieff, R.C. Mulligan, R.P. Johnson (1997) Dye efflux studies suggest that hematopoietic stem cells expressing low or undetectable levels of CD34 antigen exist in multiple species. Nature Med *3:* 1337–1345.
Gussoni, E., Y. Soneoka, C.D. Strickland, E.A. Buzney, M.K. Khan, A.F. Flint, L.M. Kunkel, R.C. Mulligan (1999) Dystrophin expression in the mdx mouse restored by stem cell transplantation. Nature *401:* 390–394.
Hadorn, E. (1968) Transdetermination in cells. Sci Am *219:* 110–114.
Harrison, D.E., C.M. Astle, M. Stone (1989) Numbers and functions of transplantable primitive immunohematopoietic stem cells. Effects of age. J Immunol *142:* 3833–3840.
Haskill, J.S., M.A.S. Moore (1970) Two dimensional cell separation: Comparison of embryonic and adult haemopoietic stem cells. Nature *226:* 853–854.
Hollander, G.A., B. Wang, A. Nichogiannopoulou, P.P. Platenburg, W. van Ewijk, S.J. Burakoff, J.-C. Gutierrez-Ramos, C. Terhorst (1995) Developmental control point in induction of thymic cortex regulated by a subpopulation of prothymocytes. Nature *373:* 350–353.
Huang, S., P. Law, K. Francis, B.O. Palsson, A.D. Ho (1999) Symmetry of initial cell divisions among primitive hematopoietic progenitors is independent of ontogenic age and regulatory molecules. Blood *94:* 2595–2604.
Ikuta, K., T. Kina, I. MacNeil, N. Uchida, B. Peault, Y.-H. Chien, I.L. Weissman (1990) A developmental switch in thymic lymphocyte maturation potential occurs at the level of hematopoietic stem cells. Cell *62:* 863–874.
Jacobson, L.O., E.L. Simmons, E.K. Marks, E.O. Gaston, M.J. Robson, J.H. Eldredge (1951) Further studies on recovery from radiation injury. J Lab and Clin Med *37:* 683–697.
Jordan, C.T., C.M. Astle, J. Zawadzki, K. Mackarehtschian, I.R. Lemischka, D.E. Harrison (1995) Long-term repopulating abilities of enriched fetal liver stem cells measured by competitive repopulation. Exp Hematol *23:* 1011–1015.
Jordan, C.T., J.P. McKearn, I.R. Lemischka (1990) Cellular and developmental properties of fetal hematopoietic stem cells. Cell *61:* 953–963.
Keller, G., R. Snodgrass (1990) Life span of multipotential hematopoietic stem cells in vivo. J Exp Med *171:* 1407–1418.
Klinken, S.P., W.S. Alexander, J.M. Adams (1988) Hemopoietic lineage switch: v-raf oncogene converts Eλ-myc transgenic B cells into macrophages. Cell *53:* 857–867.
Kondo, M., K. Akashi, J. Domen, K. Sugamura, I.L. Weissman (1997) Bcl-2 rescues T lymphopoiesis, but not B or NK cell development, in common gamma chain-deficient mice. Immunity *7:* 155–162.
Kondo, T., M. Raff (2000) Oligodendrocyte precursor cells reprogrammed to become multipotential CNS stem cells. Science *289:* 1754–1757.
Krause, D.S., N.D. Theise, M.I. Collector, O. Henegariu, S. Hwang, R. Gardner, S. Neutzel, S.J. Sharkis (2001) Multi-organ, multi-lineage engraftment by a single bone marrow-derived stem cell. Cell *105:* 369–377.
Lagasse, E., H. Connors, M. Al-Dhalimy, M. Reitsma, M. Dohse, L. Osborne, X. Wang, M. Finegold, I.L. Weissman, M. Grompe (2000) Purified hematopoietic stem cells can differentiate into hepatocytes in vivo. Nat Med *6:* 1229–1234.
Loeffler, M., C.S. Potten (1997) Stem cells and cellular pedigrees – a conceptual introduction; in Potten, C.S. (ed): Stem Cells. London, Academic Press, pp 1–27.
Lois, C., A. Alvarez-Buylla (1994) Long-distance neuronal migration in the adult mammalian brain. Science *264:* 1145–1148.
Lord, B.I., E. Spooncer (1986) Isolation of haemopoietic spleen colony forming cells. Lymphokine Res *5:* 59–72.
Lorenz, E., D. Uphoff, T.R. Reid, E. Shelton (1951) Modification of irradiation injury in mice and guinea pigs by bone marrow injections. J Nat Cancer Inst *12:* 197–201.
Magli, M.C., C. Largman, H.J. Lawrence (1997) Effects of HOX homeobox genes in blood cell differentiation. J Cell Physiol *173:* 168–177.
Medvinsky, A.L. O.I. Gan, M.L. Semenova, N.L. Samoylina (1996) Development of day-8 colony-forming unit – spleen hematopoietic progenitors during early murine embryogensis: spatial and temporal mapping. Blood *87:* 557–566.
Medvinsky, A.L., N.L. Samoylina, A.M. Müller, E.A. Dzierzak (1993) An early preliver intraembryonic source of CFU-S in the developing mouse. Nature *364:* 64–67.
Mezey, E., K.J. Chandross, G. Harta, R.A. Maki, S.R. McKercher (2000) Turning blood into brain: cells bearing neuronal antigens generated in vivo from bone marrow. Science *290:* 1779–1782.
Morrison, S.J., I.L. Weissman (1994) The long-term repopulating subset of hematopoietic stem cells is deterministic and isolatable by phenotype. Immunity *1:* 661–673.
Morshead, C.M., B.A. Reynolds, C.G. Craig, M.W. McBurney, W.A. Staines, D. Morassutti, S. Weiss, D. van der Kooy (1994) Neural stem cells in the adult mammalian forebrain: a relatively quiescent subpopulation of subependymal cells. Neuron *13:* 1071–1082.
Müller, A.M., E.A. Dzierzak (1993) ES cells have only a limited lymphopoietic potential after adoptive transfer into mouse recipients. Development *118:* 1343–1351.
Müller, A.M., R. Henschler, S. Kreutzfeldt, S. Petrovic, B. Mühl, F. Harder (2000) Chimerism in embryos and adults following the injection of mouse and human hematopoietic and mouse neural stem cells into blastocysts. Blood *96:* 276a (abstract No. 1184).
Müller, A.M., A. Medvinsky, J. Strouboulis, F. Grosveld, E. Dzierzak (1994) Development of hematopoietic stem cell activity in the mouse embryo. Immunity *1:* 291–301.
Okada, S., H. Nakauchi, K. Nagayoshi, S.-I. Nishikawa, Y. Miura, T. Suda (1992) In vivo and in vitro stem cell function of c-kit- and Sca-1-positive murine hematopoietic cells. Blood *80:* 3044–3050.
Pesce, M., K. Anastassiadis, H.R. Scholer (1999) Oct-4: lessons of totipotency from embryonic stem cells. Cells Tissues Organs *165:* 144–152.
Price, J., D. Turner, C. Cepko (1987) Lineage analysis in the vertebrate nervous system by retrovirus-mediated gene transfer. Proc Natl Acad Sci USA *84:* 156–160.
Pincipato, M., J.L. Cleveland, U.R. Rapp, K.L. Holmes, J.H. Pierce, H.C. Morse 3rd, S.P. Klinken (1990) Transformation of murine bone marrow cells with combined v-raf-v-myc oncogenes yields clonally related mature B cells and macrophages. Mol Cell Biol *10:* 3562–3568.
Qian, X., S.K. Goderie, Q. Shen, J.H. Stern, S. Temple (1998) Intrinsic programs of patterned cell lineages in isolated vertebrate CNS ventricular zone cells. Development *125:* 3143–3152.
Rebel, V.I., C.L. Miller, G.R. Thornbury, W.H. Dragowska, C.J. Eaves, P.M. Lansdorp (1996) A comparison of long-term repopulating hematopoietic stem cells in fetal liver and adult bone marrow from the mouse. Exp Hematol *24:* 638–648.
Reid, C.B., I. Liang, C. Walsh (1995) Systematic widespread clonal organization in cerebral cortex. Neuron *15:* 299–310.
Reynolds, B.A. S. Weiss (1992) Generation of neurons and astrocytes from isolated cells of the adult mammalian central nervous system. Science *255:* 1707–1710.
Reynolds, B.A., S. Weiss (1996) Clonal and population analyses demonstrate that an EGF-responsive mammalian embryonic CNS precursor is a stem cell. Dev Biol *175:* 1–13.
Schneuwly, S., R. Klemenz, W.J. Gehring (1987) Redesigning the body plan of Drosophila by ectopic expression of the homoeotic gene Antennapedia. Nature *325:* 816–818.
Spangrude, G.J., S. Heimfeld, I.L. Weissman (1988) Purification and characterization of mouse hematopoietic stem cells. Science *241:* 58–62.
Strouboulis, J., N. Dillon, F. Grosveld (1992) Developmental regulation of a complete 70-kb human beta-globin locus in transgenic mice. Genes Dev *6:* 1857–1864.
Suhonen, J.O., D.A. Peterson, J. Ray, F.H. Gage (1996) Differentiation of adult hippocampus-derived progenitors into olfactory neurons in vivo. Nature *383:* 624–627.

Szilvassy, S.J., R.K. Humphries, P.M. Lansdorp, A.C. Eaves, C.J. Eaves (1990) Quantitative assay for totipotent reconstituting hematopoietic stem cells by a competitive repopulation strategy. Proc Natl Acad Sci USA *87:* 8736–8740.

Traycoff, C.M., A. Orazi, A.C. Ladd, S. Rice, J. McMahel, E.F. Srour (1998) Proliferation-induced decline of primitive hematopoietic progenitor cell activity is coupled with an increase in apoptosis of ex vivo expanded CD34+ cells. Exp Hematol *26:* 53–62.

Trevisan, M., N.N. Iscove (1995) Phenotypic analysis of murine long-term hemopoietic reconstituting cells quantitated competitively in vivo and comparison with more advanced colony-forming progeny. J Exp Med *181:* 93–103.

Turner, R.W.A., L. Siminovitch, E.A. McCulloch, J.E. Till (1966) Density gradient centrifugation of hemopoietic colony-forming cells. J Cell Physiol *69:* 73–82.

Uchida, N., I.L. Weissman (1992) Searching for hematopoietic stem cells: evidence that Thy-1.1 lo Lin^- $Sca\text{-}1^+$ cells are the only stem cells in C57BL/Ka-Thy-1.1 bone marrow. J Exp Med *175:* 175–184.

Vescovi, A.L., B.A. Reynolds, D.D. Fraser, S. Weiss (1993) bFGF regulates the proliferative fate of unipotent (neuronal) and bipotent (neuronal/astroglial) EGF-generated CNS progenitor cells. Neuron *11:* 951–966.

Vescovi, A.L., E.Y. Snyder (1999) Establishment and properties of neural stem cell clones: plasticity in vitro and in vivo. Brain Pathol *9:* 569–598.

Visser, J.W.M., J.G.J. Bauman, A.H. Mulder, J.F. Eliason, A.M. de Leeuw (1984) Isolation of murine pluripotent hemopoietic stem cells. J Exp Med *159:* 1576–1590.

Watt, F.M., B.L.M. Hogan (2000) Out of eden: stem cells and their niches. Science *287:* 1427–1430.

Wei, G., G. Schubiger, F. Harder, A.M. Müller (2000) Stem cell plasticity in mammals and transdetermination in Drosophila: common themes? Stem Cells *18:* 409–414.

Weiss, S., C. Dunne, J. Hewson, C. Wohl, M. Wheatley, A.C. Peterson, B.A. Reynolds (1996) Multipotent CNS stem cells are present in the adult mammalian spinal cord and ventricular neuroaxis. J Neurosci *16:* 7599–7609.

Whetton, A.D. G.J. Graham (1999) Homing and mobilization in the stem cell niche. Trends Cell Biol *9:* 233–238.

Williams, B.P., J. Price (1995) Evidence for multiple precursor cell types in the embryonic rat cerebral cortex. Neuron *14:* 1181–1188.

Wu, A.M., J.E. Till, L. Siminovitch, E.A. McCulloch (1968) Cytological evidence for a relationship between normal hematopoietic colony-forming cells and cells of the lymphoid system. J Exp Med *127:* 455–463.

Zhong, R.-K., C.M. Astle, D.E. Harrison (1996) Distinct developmental patterns of short-term and long-term functioning lymphoid and myeloid precursors defined by competitive limiting dilution analysis in vivo. J Immunol *157:* 138–145.

Cells
Tissues
Organs

Cells Tissues Organs 2002;171:90–95

Human Mesenchymal Stem Cells: Insights from a Surrogate in vivo Assay System

Tippi C. MacKenzie   Alan W. Flake

The Children's Institute for Surgical Science, The Children's Hospital of Philadelphia, Pa., USA

Key Words

Mesenchymal stem cells · Human-sheep model · In utero stem cell transplantation

Abstract

Mesenchymal stem cells (MSC) are multipotent cells that have been isolated from the bone marrow of multiple species that can be induced to differentiate into at least bone, cartilage, and adipose tissue in vitro. Using a model of in utero cellular transplantation in which human MSC are transplanted into fetal lambs, we have shown that MSC can engraft in multiple tissues and persist for over one year. Furthermore, these cells can differentiate into cardiac and skeletal myocytes, bone marrow stromal cells, adipocytes, thymic epithelial cells, and chondrocytes. These observations lend support to the potential utility of MSC for cellular and/or gene therapy in the treatment of a variety of congenital or acquired diseases such as osteogenesis imperfecta, muscular dystrophy, lysosomal storage diseases, and for enhancement of bone marrow transplantation.

Abbreviations used in this paper

MSC	mesenchymal stem cells
HSC	hematopoietic stem cells
BM	bone marrow
PCR	polymerase chain reaction
OI	osteogenesis imperfecta

Introduction

Mesenchymal stem cells (MSC) are of potential interest to investigators in the fields of tissue engineering, gene therapy and cellular transplantation. MSC, as currently defined, are a population of multipotent cells that can be isolated from the bone marrow by plastic adherence, and can differentiate in vitro into mesenchymal tissues [Caplan, 1991, 1994]. Cell populations containing MSC have been successfully isolated from various species, expanded in culture, and demonstrated to have the capacity to differentiate into bone, cartilage, and adipose tissue in vitro [Prockop, 1997]. It has become increasingly apparent that characterization of MSC is a process in evolution, and that current populations represent a heterogeneous mixture of cells that include only a variably enriched subpopulation of stem cells. Further characterization of MSC awaits the development of specific phenotypic and functional markers, and demonstration of multipotentiality in both in vitro and in vivo assay systems at a clonal level.

KARGER

Fax + 41 61 306 12 34
E-Mail karger@karger.ch
www.karger.com

1422–6405/02/1711–0090$18.50/0

Accessible online at:
www.karger.com/journals/cto

Alan W. Flake
The Children s Institute for Surgical Science, The Children s Hospital of Philadelphia
34th Street and Civic Center Boulevard
Philadelphia, PA 19104 (USA)
Tel. +1 215 590 3188, Fax +1 215 590 3324, E-Mail Flake@email.chop.edu

In vitro Multipotentiality of MSC

Human MSC from bone marrow aspirates have been isolated and induced to show differentiation into osteocytes, chondrocytes, and adipocytes in vitro [Pittenger et al., 1999]. Muscle differentiation of MSC has also been reported under specific culture conditions with 5-azacytidine [Wakitani et al., 1995]. More recently, endothelial cell differentiation of MSC has been reported in vitro after culturing the cells with vascular endothelial growth factor [Reyes and Verfaillie, 2000a]. MSC may also have the ability to differentiate into neurons. Woodbury et. al. [2000] reported neuronal differentiation of rat and human MSC in specific culture conditions, while Reyes and Verfaillie [1999, 2000b] have also shown both astrocytic and neuronal differentiation of human MSC in vitro. It is important to emphasize that the expression of transcription factors and surface markers of lineage differentiation in culture does not represent proof of functionality of these cells. It is clear that both the isolation as well as the culture conditions contribute to the realization of their plasticity. The physiologic role of MSC needs to be determined in in vivo assay systems. The development of more refined stem cell populations will enhance our ability to analyze the true functionality of these cells and to utilize them clinically.

In vivo Multipotentiality of MSC after in utero Transplantation

Although in vitro data supports the potential utility of MSC for cellular or gene therapy, there is currently minimal in vivo data on the transplantation biology or differentiation potential of these cells after systemic administration. We have recently utilized a human-sheep xenogeneic transplantation model in order to explore the in vivo engraftment and differentiation potential of MSC after systemic administration into the fetal environment. This model was originally developed as a surrogate assay system for human hematopoesis. Transplantation of human hematopoetic stem cells (HSC) into pre-immune fetal lambs resulted in secondary multilineage engraftment and long-term mixed chimerism [Zanjani et al., 1992, 1994]. Since the early gestational fetal lamb is tolerant to xenogeneic cells, and since transplanted human cells may be distinguished from host cells with a variety of methodologies, we reasoned that the model would be useful for the assay of human MSC as well. Furthermore, the fetal environment may provide unique inductive signals to support the rapid proliferation and differentiation of transplanted stem cells.

In order to study the homing, engraftment, and differentiation of MSC after systemic administration, we injected a relatively defined population of human MSC [Pittenger et al., 1999] into both pre- and post-immune fetal lambs (65 and 85 days of gestation, respectively). Fetal lambs received intraperitoneal injections of human MSC and were sacrificed at several time points after transplantation, up to 13 months of age. We then used PCR, immunohistochemistry for β-2 microglobulin (a component of Class I MHC), and in situ hybridization for human Alu sequences to detect human cells in various tissues.

Overall, 95% of the animals had evidence of engraftment in one or more tissues [Liechty et al., 2000]. Early time points (2 weeks) showed engraftment in almost all tissues tested (blood, bone marrow, brain, liver, spleen, muscle, heart, and thymus) by PCR, whereas the cells survived in fewer tissues in later time points. PCR-positive tissues showed evidence of human cells after staining with β-2 microglobulin (fig. 1a) or with in situ hybridization. We also studied whether these MSCs underwent site-specific differentiation by staining either with human-specific markers of differentiation, or by double staining for a human marker and a marker for tissue differentiation. With these methods, we observed differentiation into cardiac muscle cells (by staining for SERCA-2, a component of skeletal muscle, and in situ hybridization) (fig. 1b), skeletal muscle cells (by double staining with myosin and β-2 microglobulin) (fig. 1c), bone marrow stroma (by staining with human-specific CD23), and thymic epithelial cells (by staining with CD74) (fig. 1d). Human-derived cells were also observed in lung, brain, spleen, and cartilage, further indicating the broad distribution of these cells.

These results pose interesting new questions for further inquiry. For example, MSCs may be immunologically privileged cells. Although the 85 day gestation fetal lamb is capable of rejecting allografts [Silverstein et al., 1964], we saw no difference in engraftment or persistence between animals who received cells at 65 or 85 days. MSC may have an immunomodulatory role in vivo [McIntosh et al., 1999]. One possibility is that MSC induce T cell anergy by expression of HLA Class I antigen in the absence of co-stimulatory signals. MSC have low or no expression of HLA Class II molecules and do not express costimulatory receptors or ligands. In addition, MSC suppress proliferative responses to allogeneic or xenogeneic antigens in mixed lymphocyte reactions by an as yet undetermined mechanism [Klyushnenkova et al., 1998].

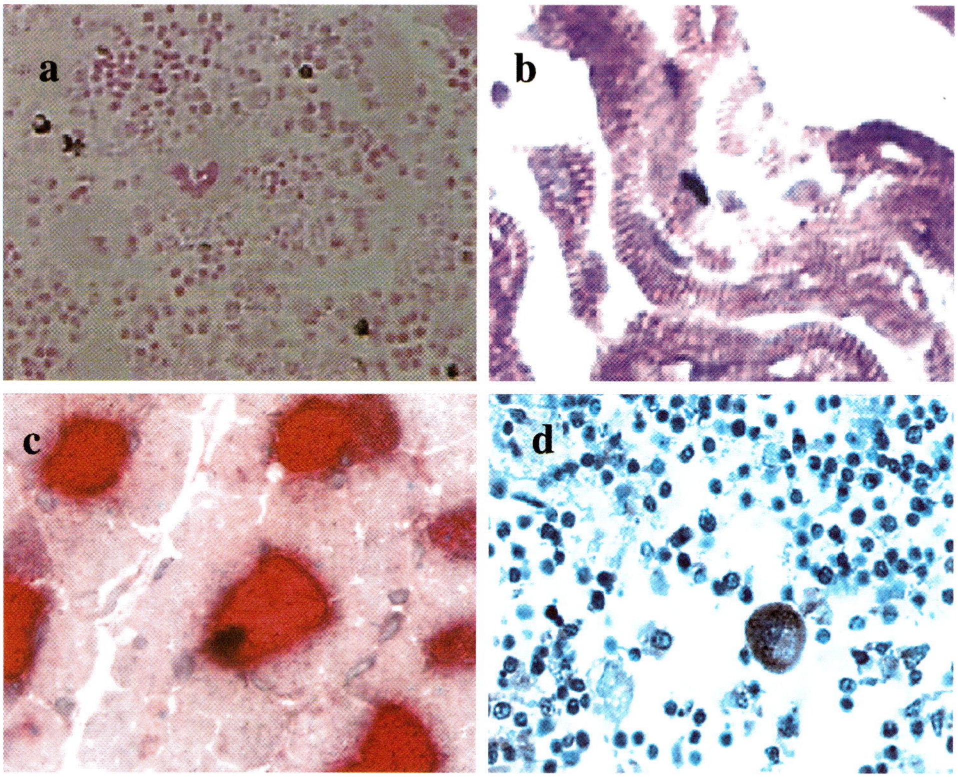

Fig. 1. Immunohistochemistry on chimeric sheep tissues demonstrating the presence of engrafted human cells with site-specific differentiation. Adapted from Liechty et al. [2000]. **a** Chimeric bone marrow stained with human-specific β-2 microglobulin (brown), demonstrating multiple human-derived cells (× 40). **b** Chimeric cardiac muscle after in situ hybridization for human Alu sequence (black) and immunohistochemistry for SERCA-2 (pink) demonstrating one human-derived cardiomyocyte (× 100). **c** Chimeric skeletal muscle stained with myosin heavy chain (slow type, red) and β-2 microglobulin (black), demonstrating one human-derived skeletal myocyte (× 100). **d** Chimeric thymus stained with human-specfic CD74 (brown), demonstrating one human-derived thymic epithelial cell (× 100).

Another group has shown similar distribution by PCR of a different population of human MSC transplanted into fetal lamb recipients, as well as differentiation of these cells into hematopoetic cells and hepatocytes [Almeida-Porada et al., 2000a]. This illustrates that the fetal environment is particularly conducive to promoting the widespread engraftment and differentiation of transplanted cells and also suggests that the differentiative capacity of MSC will be dependent on not only the assay system utilized but also the method of isolation. In the future, other MSC populations will need to be comparatively assayed.

Although the in vivo differentiation capacity of MSC demonstrated in the fetal sheep model is supportive of their potential clinical utility, realistic strategies for the utilization of MSC for cellular or gene therapy will require the ability to deliver a large number of MSC to specific sites, or the ability to regulate the homing, differentiation and proliferation to achieve therapeutically significant numbers of MSC in the target tissues.

Potential Applications of MSC Transplantation

Much of the enthusiasm regarding the potential of MSC derives from their ready availability via bone marrow biopsy, and the relative simplicity of their expansion without loss of multipotentiality. These features make current MSC populations promising for applications involving tissue replacement or repair, such as bone or cartilage generation or for tissue engineering. Autologous cells could also potentially be used for gene therapy applications requiring secretion of a protein. However, immune reaction to the encoded gene product with consequent silencing may still be a problem. Furthermore, for many diseases such as muscular dystrophy, an allogeneic source will need to be used, which would require immunosuppressive conditioning. To avoid the immune response and for many other reasons, we have been interested in the prenatal transplantation of MSC as a therapeutic strategy for specific disorders.

There are many potential advantages of prenatal cellular or gene therapy over postnatal treatment approaches. Experience with HSC transplantation has shown that the fetus can accept allogeneic and even xenogeneic grafts without the need for conditioning regimens [Billingham et al., 1953]. If tolerance is induced to the allografts, post-

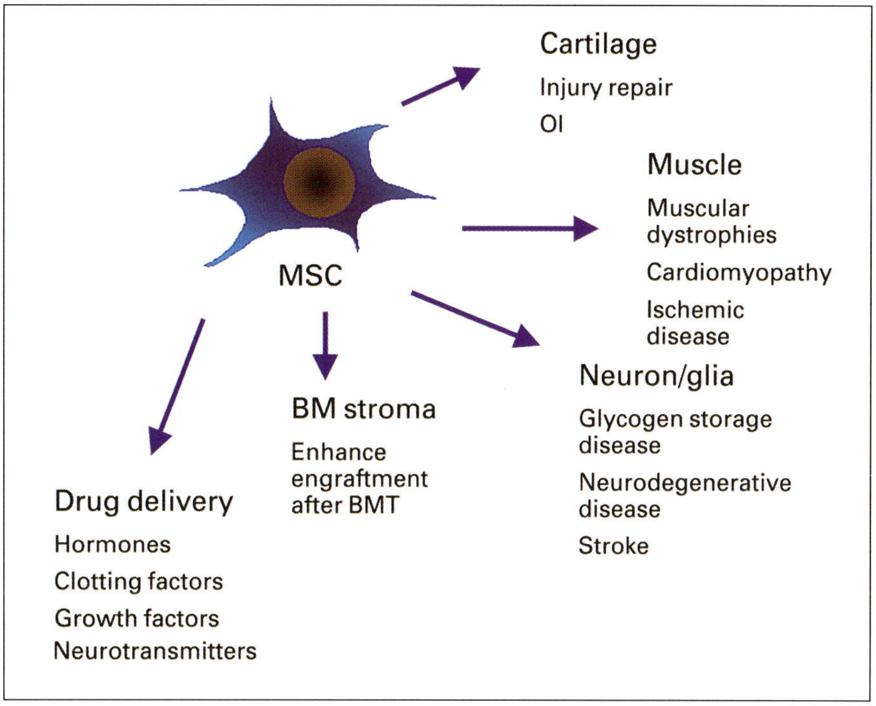

Fig. 2. Clinical applications of cellular or gene therapy with MSC.

natal same-donor transplants can be performed utilizing non-myeloablative regimens to increase engraftment. Induction of tolerance to the transgene may overcome some of the current barriers to long-term expression in vivo. Moreover, the highly proliferative microenvironment of the fetus may be particularly conducive to the widespread engraftment and differentiation of transplanted stem cells. This may be especially pronounced if genetically modified cells have a survival advantage over host cells. As prenatal diagnosis and treatment becomes more routine, in utero cellular and gene therapy may be used to treat and preempt postnatal complications of congenital diseases.

Numerous congenital and acquired diseases could potentially benefit from MSC transplantation (fig. 2). One of the first examples of the clinical utility of MSC transplantation has been for osteogenesis imperfecta (OI). Mice with OI have been transplanted with murine MSC with resultant contributions to bone and cartilage formation from the transplanted cells [Pereira et al., 1998]. Furthermore, patients with OI who have received bone marrow transplants have shown clinical improvement [Horwitz et al., 1999, 2001]. Although a pure population of MSCs was not transplanted, it is likely that a MSC population in the transplanted bone marrow is responsible for this finding.

The idea that myogenic precursors may be found in the bone marrow originated from two important experiments tracking bone marrow-derived cells into injured muscles [Ferrari et al., 1998] or mdx muscles [Bittner et al., 1999] following bone marrow transplantation. Purified hematopoetic stem cells have been shown to migrate into areas of injury and differentiate into muscle in the mdx mouse [Gussoni et al., 1999]. Locally delivered c-kit+ stem cells injected into the vicinity of infarcted myocardium were recently shown to differentiate into cardiomyocytes [Orlic et al., 2001]. In our experiments, systemically delivered human MSC differentiated into skeletal and cardiac myocytes in lambs. Whether myogenic stem cells arise from MSC or HSC, cellular transplantation will clearly be an important tool for treatment of muscular dystrophy in the future. Given the need to achieve engraftment in multiple muscle groups, and the immunologic issues that have encumbered myoblast transplantation efforts [Miller et al., 1997], a prenatal approach may be advantageous.

Lysosomal storage diseases are another exciting frontier for MSC transplantation. Multiple groups have demonstrated that following bone marrow transplantation, donor cells may be found in the brain and differentiate into microglia [Krall et al., 1994] and neurons [Brazelton et al., 2000]. Furthermore, there is evidence that MSC transplanted into the lateral ventricles of newborn mice migrate extensively and differentiate into astrocytes [Kopen et al., 1999]. Even if transplanted cells do not differentiate, they may be used to locally deliver missing enzymes. Our observation of MSCs engrafting in perivascular tissues within the brain after systemic injection, even if differentiation was not observed, may be important in this regard.

Finally, there is evidence that MSCs may support the engraftment of hematopoetic stem cells. In the absence of myeloablation, both pre- and postnatal bone marrow transplantation efforts have been plagued by failure of engraftment in the absence of a selective advantage for donor cells [Flake and Zanjani, 1999]. Clinical successes have been achieved in patients with SCID [Flake et al., 1996; Wengler et al., 1996], but absence of adequate levels of engraftment in other circumstances prevents widespread use of this method for diseases such as hemoglobinopathies and inborn errors of metabolism. Since MSC differentiate into bone marrow stroma in vivo and support the proliferation of hematopoetic stem cells in vitro [Majumdar et al., 1998], they may provide a stromal microenvironment that can promote the engraftment of donor cells following bone marrow transplantation. This hypothesis has been proven in a sheep model with improved levels of engraftment and earlier expression of donor cells after in utero transplantation of HSC with bone marrow stroma [Almeida-Porada et al., 1999, 2000b].

Gene therapy applications of MSC are particularly promising. MSC from different species can be transduced with retroviruses [Mosca et al., 2000] and retain their expression in vivo. Furthermore, transduced MSC may be implanted [Allay et al., 1997] and retain their ability to differentiate [Marx et al., 1999]. MSC expressing bone-morphogenic protein (BMP)-2 [Gazit et al., 1999] and BMP-7 [Mason et al., 2000] have been shown to contribute to bone formation in animal models of repair. Given these encouraging results, genetically engineered MSC may also be used to deliver hormones, clotting factors, and other proteins systemically. Tissue-specific and inducible promoters may be designed to achieve more controlled delivery.

Conclusions

At the present time, MSC represent poorly defined heterogeneous populations of cells that differ in their in vitro and in vivo potentiality. A more refined understanding of the biologic potential and normal physiologic role of MSC awaits the development of better isolation procedures and specific cell surface and functional markers as well as data from in vivo assay systems. Exploitation of the full clinical potential of MSC will require a better understanding of their normal physiological function and the regulatory signals controlling their homing, proliferation, and differentiation. Nevertheless, early data is promising that MSC may have important clinical applications for diseases such as OI and DMD and that genetically engineered MSC may extend the disease targets beyond those of mesenchyme-derived tissues.

An important extension of our studies is the development of a fetal allogeneic model of MSC transplantation. One model could be the mouse, in which it is also possible to perform in utero injections during the pre-immune period and show engraftment of allogeneic HSC [Shaaban et al., 1999]. This model would avoid perturbations of species specificity and allow testing of the immunologic aspects of MSC transplantation as well as investigations in various established transgenic mice. In utero injections may be site-directed to test whether locally injected cells survive as well as systemically injected cells. Although murine MSC have been difficult to isolate and characterize given differences in yield among strains [Phinney et al., 1999], we and other groups have recently made progress in this area [Kitano et al., 2000]. Regardless of the model used, further understanding of MSC biology will be critically dependent upon validation in in vivo systems. The documentation of functional integration of these cells will be essential prior to designing clinical strategies for cellular therapy.

References

Allay, J.A., J.E. Dennis, S.E. Haynesworth, M.K. Majumdar, D.W. Clapp, L.D. Shultz, A.I. Caplan, S.L. Gerson (1997). LacZ and interleukin-3 expression in vivo after retroviral transduction of marrow-derived human osteogenic mesenchymal progenitors. Hum Gene Ther *8:* 1417–1427.

Almeida-Porada, G., A.W. Flake, H.A. Glimp, E.D. Zanjani (1999). Cotransplantation of stroma results in enhancement of engraftment and early expression of donor hematopoietic stem cells in utero. Exp Hematol *27:* 1569–1575.

Almeida-Porada G., C. Porada, N. Brouard, P.J. Simmons, J.L.Ascensao, E.D. Zanjani (2000a). Generation of hematopoetic and hepatic cells by human bone marrow stromal cells in vivo. Blood *96:* 570a (abstract 2451).

Almeida-Porada, G., C.D. Porada, N. Tran, E.D. Zanjani (2000b). Cotransplantation of human stromal cell progenitors into preimmune fetal sheep results in early appearance of human donor cells in circulation and boosts cell levels in bone marrow at later time points after transplantation. Blood *95:* 3620–3627.

Billingham R.E., L.Brent, P.B. Medawar (1953). 'Actively acquired tolerance' of foreign cells. Nature *172:* 603–606.

Bittner, R.E., C. Schofer, K. Weipoltshammer, S. Ivanova, B. Streubel, E. Hauser, M. Freilinger, H. Hoger, A. Elbe-Burger, F. Wachtler (1999). Recruitment of bone-marrow-derived cells by skeletal and cardiac muscle in adult dystrophic mdx mice. Anat Embryol (Berl) *199:* 391–396.

Brazelton, T.R., F.M. Rossi, G.I. Keshet, H.M. Blau (2000). From marrow to brain: expression of neuronal phenotypes in adult mice. Science *290:* 1775–1779.

Caplan, A.I. (1991). Mesenchymal stem cells. J Orthop Res *9:* 641–650.

Caplan, A.I. (1994). The mesengenic process. Clin Plast Surg *21:* 429–435.

Ferrari, G., G. Cusella-De Angelis, M. Coletta, E. Paolucci, A. Stornaiuolo, G. Cossu, F. Mavilio (1998). Muscle regeneration by bone marrow-derived myogenic progenitors. Science *279:* 1528–1530.

Flake, A.W., M.G. Roncarolo, J.M. Puck, G. Almeida-Porada, M.I. Evans, M.P. Johnson, E.M. Abella, D.D. Harrison, E.D. Zanjani (1996). Treatment of X-linked severe combined immunodeficiency by in utero transplantation of paternal bone marrow. N Engl J Med *335:* 1806–1810.

Flake, A.W., E.D. Zanjani (1999). In utero hematopoietic stem cell transplantation: ontogenic opportunities and biologic barriers. Blood *94:* 2179–2191.

Gazit, D., G. Turgeman, P. Kelley, E. Wang, M. Jalenak, Y. Zilberman, I. Moutsatsos (1999). Engineered pluripotent mesenchymal cells integrate and differentiate in regenerating bone: a novel cell-mediated gene therapy. J Gene Med *1:* 121–133.

Gussoni, E., Y. Soneoka, C.D. Strickland, E.A. Buzney, M.K. Khan, A.F. Flint, L.M. Kunkel, R.C. Mulligan (1999). Dystrophin expression in the mdx mouse restored by stem cell transplantation. Nature *401:* 390–394.

Horwitz, E.M., D.J. Prockop, L.A. Fitzpatrick, W.W. Koo, P.L. Gordon, M. Neel, M. Sussman, P. Orchard, J.C. Marx, R.E. Pyeritz, M.K. Brenner (1999). Transplantability and therapeutic effects of bone marrow-derived mesenchymal cells in children with osteogenesis imperfecta. Nat Med *5:* 309–313.

Horwitz, E.M., D.J. Prockop, P.L. Gordon, W.W. Koo, L.A. Fitzpatrick, M.D. Neel, M.E. McCarville, P.J. Orchard, R.E. Pyeritz, M.K. Brenner (2001). Clinical responses to bone marrow transplantation in children with severe osteogenesis imperfecta. Blood *97:* 1227–1231.

Kitano, Y., A. Radu, A. Shaban, A.W. Flake (2000). Selection, enrichment, and culture expansion of murine mesenchymal progenitor cells by retroviral transduction of cycling adherent bone marrow cells. Exp Hematol *28:* 1460–1469.

Klyushnenkova, E., J.D. Mosca, K.R. McIntosh (1998). Human mesenchymal stem cells suppress allogeneic T cell responses in vitro: implications for allogeneic transplantation. Blood *92:* 642a.

Kopen, G.C., D.J. Prockop, D.G. Phinney (1999). Marrow stromal cells migrate throughout forebrain and cerebellum, and they differentiate into astrocytes after injection into neonatal mouse brains. Proc Natl Acad Sci USA *96:* 10711–10716.

Krall, W.J., P.M. Challita, L.S. Perlmutter, D.C. Skelton, D.B. Kohn (1994). Cells expressing human glucocerebrosidase from a retroviral vector repopulate macrophages and central nervous system microglia after murine bone marrow transplantation. Blood *83:* 2737–2748.

Liechty, K.W., T.C. MacKenzie, A.F. Shaaban, A. Radu, A.B. Moseley, R. Deans, D.R. Marshak, A.W. Flake (2000). Human mesenchymal stem cells engraft and demonstrate site-specific differentiation after in utero transplantation in sheep. Nat Med *6:* 1282–1286.

Majumdar, M.K., M.A. Thiede, J.D. Mosca, M. Moorman, S.L. Gerson (1998). Phenotypic and functional comparison of cultures of marrow-derived mesenchymal stem cells (MSCs) and stromal cells. J Cell Physiol *176:* 57–66.

Marx, J.C., J.A. Allay, D.A. Persons, S.A. Nooner, P.W. Hargrove, P.F. Kelly, E.F. Vanin, E.M. Horwitz (1999). High-efficiency transduction and long-term gene expression with a murine stem cell retroviral vector encoding the green fluorescent protein in human marrow stromal cells. Hum Gene Ther *10:* 1163–1173.

Mason, J.M., A.S. Breitbart, M. Barcia, D. Porti, R.G. Pergolizzi, D.A. Grande (2000). Cartilage and bone regeneration using gene-enhanced tissue engineering. Clin Orthop *54:* S171–178.

McIntosh, K., E. Klyushnenkova, V. Shustova, A. Moseley, R. Deans (1999). Suppression of alloreactive T cell responses by human mesenchymal stem cells involves CD8+ cells. Blood *94:* 133a (abstract 587).

Miller, R.G., K.R. Sharma, G.K. Pavlath, E. Gussoni, M. Mynhier, A.M. Lanctot, C.M. Greco, L. Steinman, H.M. Blau (1997). Myoblast implantation in Duchenne muscular dystrophy: the San Francisco study. Muscle Nerve *20:* 469–478.

Mosca, J.D., J.K. Hendricks, D. Buyaner, J. Davis-Sproul, L.C. Chuang, M.K. Majumdar, R. Chopra, F. Barry, M. Murphy, M.A. Thiede, U. Junker, R.J. Rigg, S.P. Forestell, E. Bohnlein, R. Storb, B.M. Sandmaier (2000). Mesenchymal stem cells as vehicles for gene delivery. Clin Orthop *54:* S71–90.

Orlic, D., J. Kajstura, S. Chimenti, I. Jakoniuk, S.M. Anderson, B. Li, J. Pickel, R. McKay, B. Nadal-Ginard, D.M. Bodine, A. Leri, P. Anversa (2001). Bone marrow cells regenerate infarcted myocardium. Nature *410:* 701–705.

Pereira, R.F., M.D. O'Hara, A.V. Laptev, K.W. Halford, M.D. Pollard, R. Class, D. Simon, K. Livezey, D.J. Prockop (1998). Marrow stromal cells as a source of progenitor cells for nonhematopoietic tissues in transgenic mice with a phenotype of osteogenesis imperfecta. Proc Natl Acad Sci USA *95:* 1142–1147.

Phinney, D.G., G. Kopen, R.L. Isaacson, D.J. Prockop (1999). Plastic adherent stromal cells from the bone marrow of commonly used strains of inbred mice: variations in yield, growth, and differentiation. J Cell Biochem *72:* 570–585.

Pittenger, M.F., A.M. Mackay, S.C. Beck, R.K. Jaiswal, R. Douglas, J.D. Mosca, M.A. Moorman, D.W. Simonetti, S. Craig, D.R. Marshak (1999). Multilineage potential of adult human mesenchymal stem cells. Science *284:* 143–147.

Prockop, D.J. (1997). Marrow stromal cells as stem cells for nonhematopoietic tissues. Science *276:* 71–74.

Reyes, M., C. Verfaillie (1999). Turning marrow into brain: in vitro differentiation of post-natal human bone marrow stem cells into neurons, oligodendrocytes, and astrocytes. Blood *94:* 377a (abstract 1676).

Reyes, M., C.M. Verfaillie (2000a). Endothelial cells generated from human bone-marrow derived mesenchymal stem cells (MSC). Blood *96:* 530a (abstract 2276).

Reyes, M., and Verfaillie C.M. (2000b). Blood *96:* 494a (abstract 2126).

Shaaban, A.F., H.B. Kim, R. Milner, A.W. Flake (1999). A kinetic model for the homing and migration of prenatally transplanted marrow. Blood *94:* 3251–3257.

Silverstein, A.M., R.A. Prendergast, K.L. Kraner (1964). Fetal response to antigenic stimulus IV. Rejection of skin homografts by the fetal lamb. J Exp Med *119:* 955–964.

Wakitani, S., T. Saito, A.I. Caplan (1995). Myogenic cells derived from rat bone marrow mesenchymal stem cells exposed to 5-azacytidine. Muscle Nerve *18:* 1417–1426.

Wengler, G.S., A. Lanfranchi, T. Frusca, R. Verardi, A. Neva, D. Brugnoni, S. Giliani, M. Fiorini, P. Mella, F. Guandalini, E. Mazzolari, S. Pecorelli, L.D. Notarangelo, F. Porta, A.G. Ugazio (1996). In-utero transplantation of parental CD34 haematopoietic progenitor cells in a patient with X-linked severe combined immunodeficiency (SCIDXI). Lancet *348:* 1484–1487.

Woodbury, D., E.J. Schwarz, D.J. Prockop, I.B. Black (2000). Adult rat and human bone marrow stromal cells differentiate into neurons. J Neurosci Res *61:* 364–370.

Zanjani, E.D., A.W. Flake, H. Rice, M. Hedrick, M. Tavassoli (1994). Engraftment and long-term expression of human fetal hemopoietic stem cells in sheep following transplantation in utero. J Clin Invest *93:* 1051–1055.

Zanjani, E.D., M.G. Pallavicini, J.L. Ascensao, A.W. Flake, R.G. Langlois, M. Reitsma, F.R. MacKintosh, D. Stutes, M.R. Harrison, M. Tavassoli (1992). Long-term repopulating ability of xenogeneic transplanted human fetal liver hematopoietic stem cells in sheep. J Clin Invest *89:* 1178–1188.

Author Index Vol. 171, No. 1, 2002

Cells Tissues Organs

Subject Index Vol. 171, No. 1, 2002

KARGER

Fax + 41 61 306 12 34
E-Mail karger@karger.ch
www.karger.com

Accessible online at:
www.karger.com/journals/cto

Conference Calendar

21.5.–25.5.2002 **Chengdu** China	1st International Tissue Engineering Meeting of China	**Information:** orthop@public.cd.sc.cn
28.5.–1.6.2002 **Cleveland, Ohio** USA	Cell Based Therapies and Tissue Engineering	**Information:** Case Western Reserve University, USA Tel. +1 216 368 3562, Fax +1 216 368 4077
28.5.–1.6.2002 **Baltimore, Md.** USA	2002 Joint Conference of the Wound Healing Society and the European Tissue Repair Society	**Information:** Wound Healing Society, 1550 S. Coast Hwy., Ste. 201, Laguna Beach, CA 92651, USA Tel. +1 888 434 4234, www.woundheal.org
31.5.–2.6.2002 **Regensburg** Germany	5th Annual Meeting of the GZG Heterologous Cell Interactions & 3-D Cell Systems	**Information:** Universität Regensburg, Germany Tel. +49 941 9448906, Fax +49 941 9446602 E-Mail gzg.2002@klinik.uni-regensburg.de http://www.uni-regensburg.de/Fakultaeten/Medizin/Pathologie/GZGconference/
9.6.–14.6.2002 **Capri** Italy	International Conference on Advances in Biomaterials for Reconstructive Medicine	**Information:** Hotel La Residenza, Capri, Italy Tel. +39 081 768 2513, Fax +39 081 242 5929 E-Mail montanin@unina.it http://www.area.na.cnr.it/itmc/
June 2002 **Seoul** South Korea	29th International Symposium on Controlled Release of Bioactive Materials	**Information:** Controlled Release Society Inc., 1020 Milwaukee Ave, Suite 335, Deerfield, IL 60015, USA Fax +1 847 808 7073, E-Mail crs@crsadmndq.org
15.6.–18.6.2002 **Toronto** Canada	4th ICRS Symposium	**Information:** 4th ICRS Symposium, Metro Toronto Convention Center, Toronto, Canada Tel. +1 416 586 8516, Fax +1 416 586 8628 E-Mail rkandel@mtsinai.on.ca
1.7.–3.7.2002 **Davos** Switzerland	ECM III Cartilage & Joint Repair: Tutorials, Basic Research & Clinical Methods	**Information:** AO Research Institute, Sonia Wahl, Secretariat E-Mail sonia.wahl@ao-aisf.ch www.ao-asif.ch/events/ao/ecm/
July 2002 **Seattle, Wash.** USA	53rd Annual Meeting of the Histochemical Society	**Information:** HS, Univ. Washington, Neurology Dept., UW Box 356465, Seattle, WA 98195, USA Fax +1 206 764 2164 E-Mail wistahl@u.washington.edu
14.7.–19.7.2002 **Florence** Italy	6th International Conference Materials in Clinical Applications	**Information:** CIMTEC 2002 Secretariat, P.O. Box 174, I–48018 Faenza, Italy Tel. +39 0546 22461/664143, Fax +39 0546 664138/663362 or visit CIMTEC 2002 web site: www.dinamica.it/cimtec
24.7.–27.7.2002 **Rio de Janeiro** Brazil	XIV Pan-American Congress on Anatomy	**Information:** President Organizing Committee: Prof. Dr. Mauricio Moscovici, E-Mail moscovic@centroin.com.br Congress Secretariat: LK Assessoria & Promoçoes Ltda, E-Mail evento@lk.com.br
11.9.–14.9.2002 **Barcelona** Spain	17th European Conference on Biomaterials	**Information:** Grupo Pacifico c/ Maria Cubi, 4, 08006 Barcelona, Spain Tel. +34 93 238 87 77, Fax +34 93 238 74 88 E-Mail gp@pacifico-meetings.com www.esb2002.com/secre.html
11.9.–15.9.2002 **Timişoara** Romania	XVIIth International Symposium on Morphological Sciences	**Information:** Prof. Dr. med. Petru Matusz, Scientific Secretariat, Department of Anatomy, University of Medicine and Pharmacy 'Victor Babes', Piata Eftimie Murgu 2, 1900 Timişoara, Romania Fax +40 56 190 626, E-Mail matusz@umft.ro Website: www.isms.2002.dnttm.ro

KARGER
Fax + 41 61 306 12 34
E-Mail karger@karger.ch
www.karger.com

Accessible online at:
www.karger.com/journals/cto

14.9.–17.9.2002 **Lyon** France	International Society of Differentiation 12th International Conference Cancer and Development with emphasis on neurobiology and cellular microenvironment	**Information:** Package Organisation, 140, Cours Charlemagne, 69002 Lyon, France Tel. +33 4 72 77 45 50, Fax +33 4 72 77 45 77 E-Mail isdmeeting2002@package.fr http://www.package.fr/ISDmeeting2002.html
15.9.–18.9.2002 **Antwerpen** Belgium	28th European Congress of Cytology	**Information:** Technologisch Instituut, c/o Ingenieurshuis – K VIV, Desguinlei 214, 2018 Antwerpen 1, Belgium Tel. +32 3 260 08 40 (general), +32 3 260 08 61 (direct), Fax +32 3 216 06 89 E-Mail cyto@conferences.ti.kviv.be www.ti.kviv.be/conf/cyto.htm
18.9.–19.9.2002 **Glasgow** UK	Annual Conference of the Tissue and Cell Engineering Society	**Information:** Glasgow University, E-Mail bea17@keele.ac.uk http://www.TCES.org
19.9.–22.9.2002 **Ankara** Turkey	IXth International Symposium on Biomedical Science and Technology (BIOMED 2002)	**Information:** BIOMED 2002 Secretariat, Ankara University Science Faculty, Tissue Engineering & Biomaterials Laboratory, Dögol Caddesi, Tandogan, 06100 Ankara, Turkey Tel. +90 312 2126720 ext. 1096, 1170 or 1023, Fax +90 312 2232395, E-Mail biomed@ankara.edu.tr http://www.biomed2002.ankara.edu.tr/default.htm
19.9.–24.9.2002 **Bamberg** Germany	7th International Symposium on Dendritic Cells	**Information:** http://www.dc2002.de
26.9.–28.9.2002 **Jena** Germany	4. DGZ-Nachwuchswissenschaftler-Tagung Embryonale und somatische Stammzellen in Grundlagenforschung und Medizin	**Information:** Sekretariat der Deutschen Gesellschaft für Zellbiologie (DGZ), Deutsches Krebsforschungszentrum, Im Neuenheimer Feld 280, 69120 Heidelberg, Germany Tel. +49 6221 42 34 51, Fax +49 6221 42 34 52 E-Mail DGZ@DKFZ.de www.zellbiologie.de
September 2002 **Genova** Italy	ETES Workshop Standards for Keratinocytes, Chondrocytes, Bone Marrow Stromal Cells	**Information:** cancedda@cda.unige.it
13.10.–16.10.2002 **Atlanta, Ga.** USA	Third Smith & Nephew International Symposium Translating Tissue Engineering Into Products	**Information:** Tel. +1 404 385 0216 www.gtec.gatech.edu
4.11.–5.11.2002 **Miami Beach, Fla.** USA	ASTM Symposium on Tissue Engineered Medical Products (TEMPs)	**Information:** Grace Lee Picciolo Tel. +1 301 827 4755 E-Mail picciolograce@mail-me.com http://www.fda.gov/CDRH/Tisseng/temps.html
14.11.–16.11.2002 **Freiburg** Germany	3. Symposium für Gynäkologische Endokrinologie und Fortpflanzungsmedizin XVI. Jahrestreffen der Deutschen IVF-Zentren	**Information:** EMC München E-Mail company@emc-event.com
25.11.–27.11.2002 **London** UK	ETES Sponsored Conference on Tissue Engineering	**Information:** robin.m@blondmcindoe.dernon.co.uk
8.12.–10.12.2002 **Kobe** Japan	5th International Meeting of the Tissue Engineering Society International	**Information:** Tel. +81 52 744 2348, Fax +81 52 744 2352 E-Mail jste@med.nagoya-u.ac.jp
14.12.–18.12.2002 **San Francisco, Calif.** USA	42nd Annual Convent of the American Society for Cell Biology	**Information:** ASCB, 9650 Rockville Pike, Bethesda, MD 20814, USA Fax +1 301 530 7139, E-Mail enewman@ascb.org
26.3.–30.3.2003 **Bonn** Germany	26. Jahrestagung der Deutschen Gesellschaft für Zellbiologie Gemeinschaftstagung der Deutschen Gesellschaft für Zellbiologie und der Gesellschaft für Entwicklungsbiologie	**Information:** Sekretariat der Deutschen Gesellschaft für Zellbiologie (DGZ), Deutsches Krebsforschungszentrum, Im Neuenheimer Feld 280, 69120 Heidelberg, Germany Tel. +49 6221 42 34 51, Fax +49 6221 42 34 52 E-Mail DGZ@DKFZ.de www.zellbiologie.de